Gas Shield

QUESTIONS & ANSWERS

Gas Shielded Arc Welding

**N. J. Henthorne and
R. W. Chadwick**

Newnes Technical Books

Newnes Technical Books

is an imprint of the Butterworth Group

which has principal offices in

London, Sydney, Toronto, Wellington, Durban and Boston

First published 1982

© Butterworth & Co (Publishers) Ltd, 1982

British Library Cataloguing in Publication Data

Henthorne, N. J.
  Gas shielded arc welding. – (Questions & answers)
  1. Shielded metal arc welding
  I. Title   II. Chadwick, R. W.   III. Series
  671.5'212   TK4660

  ISBN 0-408-01182-3

Photoset by Butterworths Litho Preparation Department
Printed in England by Woolnough Bookbinding Ltd Wellingborough, Northants

Contents

Preface

A large percentage of welding operations carried out in the modern welding world are now done so by using the gas shielded arc welding process. This book has been produced in a question and answer form in a progressive sequence. A practical course of instruction is provided, from the deposition of straight beads of weld on mild steel to the advanced techniques, and the joining of what are usually regarded as difficult-to-weld materials such as aluminium, copper and nickel.

Sections have been included on gas shielded arc spot welding along with plasma arc welding and cutting.

For the sake of clarity, the term mild steel is used in preference to low carbon steel, and welding processes are referred to by their commonly-used name, for example CO_2 welding.

We thank the British Oxygen Company Ltd. for guidance with a number of the sketches included. Extracts from BS 3019, BS 2901 and BS 499: 1980, Welding terms and symbols, are reproduced by permission of British Standards Institution, 2 Park Street, London W1, from whom copies of the complete Standards can be obtained.

We would like to thank Mr K. Leake for his valued assistance in working through the text. Acknowledgement is also due to Mr D. Jones for his help in many ways. Finally, we thank Mr Roy Eccleshare for so carefully producing the diagrams to our requirements.

N. J. Henthorne
R. W. Chadwick

1
Principles of semi-automatic welding

What is gas shielded arc welding?

Gas shielded arc welding is a group of modern welding processes in which the molten metal passing across the arc is protected from the effects of the atmosphere by means of a shielding gas. This gas is emitted from the welding nozzle which is attached to the welding gun or torch. The composition of the gas used varies according to the welding process being used and the material being welded.

The main welding processes which may be grouped under this heading include MIG, TIG and plasma arc welding.

Why are these processes widely used?

They have two important advantages. First, fluxes are not required, which means fewer weld defects and less after-weld cleaning. Secondly, the heat source is concentrated, ensuring a faster weld-metal deposition rate and more economical use of the process consumables.

What is semi-automatic welding?

This is welding in which some of the welding variables are automatically controlled, but manual guidance is necessary.

The welding processes MIG, MAG and CO_2 are semi-

1

automatic in that the consumable electrode wire is fed automatically to the weld zone, and the arc length gap is also self-adjusting.

What are the MIG, MAG and CO_2 welding processes?

MIG stands for 'metal-inert gas', where the main constituent of the shielding gas or gas mixture is inert. Normally the only inert gases used for welding are argon or helium.

MAG stands for 'metal-active gas', where a non-inert shielding gas such as nitrogen or carbon dioxide is used. These gases actively combine with some elements.

CO_2 welding is when carbon dioxide is the shielding gas being used.

What is meant by MAGS welding?

This is metal arc gas shielded welding, a term which may be used when referring to any of the above processes, irrespective of the shielding gas being used.

What is the principle of the MIG welding process?

This is a semi-automatic welding process in which an electric arc is struck between the tip of a continuous consumable bare-wire electrode and the work to be welded. The heat of the arc melts the joint edges and electrode wire, allowing them to melt and fuse together forming a welded joint. The weld area and the electrode tip are protected from atmospheric contamination by a shielding gas issuing from the nozzle of the welding gun.

What are the principal modes of metal transfer across the arc when semi-automatic welding?

There are three distinct modes of metal transfer: spray (free flight), dip (short-circuiting arc) and globular (drop). Some welding machines are capable of producing a modified form of spray transfer known as pulsed arc welding.

What is spray transfer?

The spray mode of metal transfer occurs when relatively high welding voltages and electrode wire feed speeds are used. An arc is established between the tip of a small-diameter electrode wire and the work being welded. The bare-wire electrode is automatically fed to the weld pool from a reel at a constant feed speed rate. The high voltage and wire speed (amperage) creates a high current density on the end of the electrode wire which causes its molten tip to be drawn off (magnetic pinch effect). This metal is

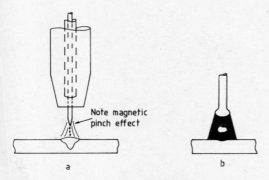

Fig. 1. (a) Spray transfer. (b) Globular transfer

then projected across the arc in the form of a very fine spray, the size of the metal droplets in the spray arc being determined by the current density on the end of the electrode wire. See Fig. 1.

With spray transfer, the arc length remains constant, and is self-adjusting even if the welding-gun-to-work distance varies slightly.

3

What is magnetic pinch effect?

This is when the high welding current passing through the small-diameter electrode wire produces a magnetic field around it which 'squeezes' its molten tip to a point.

When should spray transfer be used?

Spray transfer gives a high metal deposition rate and deep penetration. It is therefore preferred when welding sections in excess of 5 mm thickness. With this method, welding is limited to the flat position and horizontal fillet welds only, because large volumes of molten metal are difficult to control when positional welding.

However, spray transfer may be used in other welding positions on materials such as aluminium and copper. The deposited weld metal solidifies rapidly because of the high thermal conductivity of these metals.

What is dip transfer?

This is when relatively low welding voltages and electrode wire feed speeds are used. The wire feed speed will just exceed the

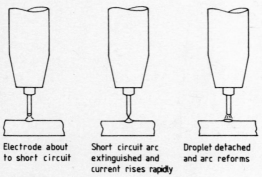

Electrode about   Short circuit arc   Droplet detached
to short circuit   extinguished and   and arc reforms
    current rises rapidly

Fig. 2. *Dip transfer*

electrode melt-off rate so that intermittent short-circuiting occurs. Each time the wire electrode tip touches the weld pool (i.e. dips into it) and short-circuits the arc, the welding current rises rapidly causing the tip of the electrode wire to be fused and detached into the weld pool (Fig. 2). The frequency of the short-circuiting arc depends upon arc volts, wire feed speed, wire diameter, welding current and the type of shielding gas being used. It may be as high as 200 per second.

When should dip transfer be used?

The heat output and metal deposition rate with this type of transfer is much less than that of spray transfer. This makes dip transfer ideally suitable for all positional welding and for the welding of thin sections.

What is pulsed MIG welding?

This is a modified form of spray transfer in which there is a periodic melting-off of droplets followed by their projection

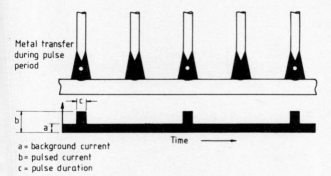

a = background current
b = pulsed current
c = pulse duration

Fig. 3. Pulsed transfer. This is a controlled spray

across the arc. In order to obtain this condition, two separate direct currents are fed to the arc: (a) a background current which keeps the gap ionized between the heated electrode tip and the weld pool, and (b) the pulse current, which is of a higher value and is applied for a brief duration at a regular frequency. See Fig. 3.

This pulse current is responsible for the melting-off and projection of the droplets across the arc at regular intervals of 50 to 100 times per second. This is a spray-type condition with a lower heat output and controlled deposition rate.

What are the applications of pulsed MIG?

The lowered heat output and reduced deposition rate allow thin sections to be welded with a spray-like condition without burn-through. Welding may now be carried out on mild steel in ALL weld positions. Compared with dip transfer welding, there is less risk of poor fusion, and spatter is almost completely eliminated, which improves the appearance of the welds.

What is globular transfer?

This mode of transfer occurs as large globules or droplets passing across the arc, the size of each globule being greater than the wire diameter. For many applications, this is undesirable because it would produce heavy spatter losses. However, globular transfer may be used successfully when welding steels with a flux-cored wire, or when welding copper with a shielding gas of nitrogen.

What is the preferred type of power source output when semi-automatic welding?

A flat characteristic (constant potential) output is preferable. With this type of power source, any slight variation in arc length causes only slight variation in the arc voltage, but the corresponding variation in the welding current is large.

What are the advantages of using this type of power source?

It permits the dip mode of metal transfer to be used by providing the necessary current surge each time the electrode short-circuits with the work. When spray transfer is being used, it provides a self-adjusting arc length condition.

What is meant by a self-adjusting arc length condition?

When using spray transfer, if the welder lowers his hand slightly and the arc length is decreased, the arc volts decrease by a small amount. This results in a large increase in the welding current

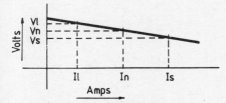

Fig. 4. *Volt/amp graph displaying flat characteristic output. Vl–Il = long arc; Vn–In = normal arc length; Vs–Is = short arc*

causing the electrode to burn back until the correct arc length is restored. The reverse occurs if the arc is lengthened. It should be noted that the wire feed speed, once adjusted, remains constant. See Fig. 4.

2

MIG, MAG and CO$_2$ welding equipment

What type of electrical power supply is necessary when using these processes?

A direct current supply is needed. This is obtained by use of a transformer rectifier welding machine which changes the a.c. mains supply to provide suitable d.c. output at the welding arc. The power source should have a flat or constant potential output characteristic. This type of machine enables the dip mode of metal transfer to be used and also provides a self-adjusting arc length condition. See page 7.

What other equipment is required?

Shielding gas supply, electrode wire feed unit, control box, regulator/flowmeter, welding gun, contact tube, nozzle, liners and conduit cable and heater for CO$_2$ welding. See Fig. 5.

How does the electrode wire feed unit operate?

It feeds the electrode wire through rollers and adjusts the pressure. The wire feed unit may be a push type, a pull type or a combined push-pull type. In the push type, the electrode wire is

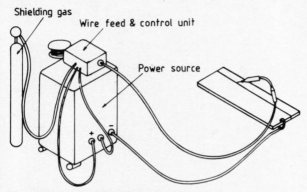

Fig. 5. Layout of CO₂ welding equipment

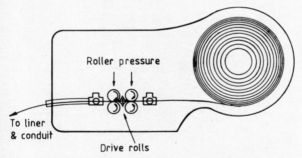

Fig. 6. Wire feed unit

pushed by the feed rollers along the conduit cable to the welding gun, the wire feed rollers being driven by a motor. See Fig. 6.

The pull type of wire feed mechanism is used for soft wires such as aluminium. A small wire spool is attached to the welding gun, and the wire feed rollers are inside the gun handle. These pull the wire through the length of the welding gun to the nozzle.

The push-pull type is a combination of the above two methods, and may be used for both hard and soft welding wires.

Which type of wire-feed unit is most suitable for feeding wires over distances over 5 m?

A combined push-pull unit is acceptable, but a helical orbiting roller drive method is also acceptable. This is known as the linear feed system, and has a linear feed motor incorporated in the conduit cable. These are placed at 5 m intervals. The reel-on gun-pull type may also be used.

What does the control box regulate?

The control box controls the wire feed speed, current on and off, shielding gas on and off, shielding gas pre- and post-purge, coolant on and off (when used). It may also be fitted with a timer to allow the MIG spot welding technique to be used.

Why is a regulator-flowmeter necessary?

The regulator reduces the gas pressure to the required working pressure as it leaves the cylinder. However, this does not guarantee a steady flow of gas, and therefore a flowmeter must be used to control and indicate the amount of gas passing to the weld area.

What influences the choice of welding gun?

The design of the welding gun is governed by: (a) the strength of current being used; (b) the type of wire feed system; and (c) access to the weld joint. When using low currents, an air-cooled goose-necked gun is preferred (Fig. 7a), whilst for the higher current ranges a water-cooled gun is available. This may be of either a goose-necked or a T shape (Fig. 7b).

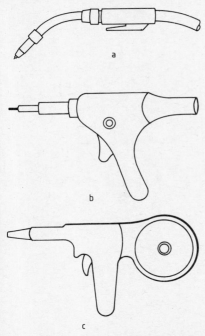

Fig. 7. Types of welding gun. (a) Air-cooled goose-necked gun. (b) Water-cooled gun suitable for welding with high current. (c) Reel-on gun type

When using the pull or the push-pull wire feed system, the welding gun will have a drive motor fixed into the handle (Fig. 7c), and a small reel of wire may be attached to the gun.

What is the purpose of the contact tube?

This is fitted into the welding gun, and supplies current to the electrode filler wire. The contact tube must be changed to accommodate different wire diameters.

11

What are the important features of the nozzle?

Replaceable nozzles are an essential part of the welding gun. These nozzles vary in diameter depending on the welding conditions and type of equipment being used. For consistently good welds, it is important that the nozzle be kept clean and free from spatter. An anti-spatter solution should be sprayed on to the nozzle periodically.

What is the purpose of the liner and conduit cable?

These are fitted between the wire feed unit and the welding gun. The flexible liner permits smooth feeding of the electrode wire to the contact tube, whilst the conduit cable contains the liner, thereby providing further protection. The conduit, along with the other services to the welding gun, is in turn enclosed in a protective rubber or plastic sheath.

Are any changes necessary when CO_2 welding?

When the shielding gas used is CO_2, a heater must be fitted between the cylinder outlet valve and the regulator.

Why is a heater required for CO_2 welding?

CO_2 is supplied in a syphon-type cylinder under a pressure of 49.3 bar, in which gaseous CO_2 is in equilibrium with liquefied CO_2. The liquid CO_2 is drawn up the syphon tube and is gasified before it leaves the cylinder. This results in a drop in temperature which the heater compensates for, thereby preventing the regulator from freezing up and thus obstructing the flow of gas to the weld area. To ensure correct working conditions, the heater should be switched on at least 5 minutes before welding is to commence. See Fig. 8.

Why is a good return lead necessary?

The return lead should be securely connected to the work or welding bench by clamping or bolting. If this is done properly, the connection may overheat and arcing between the bench and the

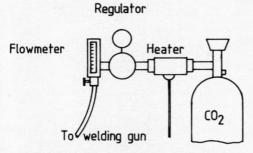

Fig. 8. CO_2 heater, flowmeter and regulator

connection could occur. It is important to remember that any increase in circuit resistance has an effect on current value, which may lead to the welding-wire electrode sticking to the work during welding.

What protective clothing should the welder wear?

Overalls, gloves, leather apron, boots, spats and leggings should be worn to give protection from heat and sparks during welding. In addition to these, a leather hood and shoulder-protectors should be worn when welding in the overhead position.

What is the purpose of the head screen?

To protect the welder's head and face from the harmful arc rays, heat and spatter losses. A filter lens of a shade suitable for the welding current being used should be fixed in the screen so that the welding arc can be seen and directed during welding. This will protect the welder from the infra-red, ultra-violet and light rays, and at the same time reduce the intensity of light to an acceptable level. A clear glass is placed in front of the filter lens to protect it from damage during welding.

A special combination lens is available for CO_2 welding which contains one 9EW lens and one 4GW lens fitted with a space between each lens.

Are any weld-cleaning accessories necessary?

A wire brush should be used to clean the weld of spatter after welding. Grinding wheels may be used to clean and prepare parts before welding, and to allow back-grinding of root runs when weld procedures specify this to be carried out.

Why should welding operations be screened?

To prevent welding arc rays causing discomfort to people working close by. To avoid reflection, walls of welding workshops should be painted with a matt finish.

Should any other hand tools be available?

Wire-cutting snips are useful to cut the electrode stick-out distance to an acceptable length when starting to weld.

3
Welding variables when MIG, MAG and CO$_2$ welding

What is meant by welding variables?

Welding variables are those factors which affect the quality of the welding arc and the weld deposit. A smooth arc condition and a high-quality weld deposit are only possible when all the welding variables are correctly adjusted. It is important that the effect of each welding variable is fully appreciated so that the full potential of the process can be achieved.

What are the important welding variables?

(1) Choice of electrode; (2) Electrode diameter; (3) Arc voltage; (4) Wire feed speed/amperage; (5) Arc length; (6) Electrode wire extension; (7) Speed of travel; (8) Shielding gas; (9) Welding gun angle; (10) Welding gun manipulation; (11) Inductance control.

Why is the selection of the correct electrode filler wire important?

The electrode wire must be compatible with the parent metal being welded. The finished weld should have the correct value of mechanical properties and a metallurgical composition similar to the parent metal. Consult manufacturers' specifications and BS 2901 1970, parts 1 to 5.

Why must the electrode wire diameter be correct?

A small-diameter wire at a fixed current setting will burn-off much faster than a larger-diameter wire at the same current value. This is because of a greater current density. As the current density on the wire increases, so the burn-off rate also increases. Small-diameter wires therefore tend to give a faster deposition rate along with deeper penetration for a given current value. Remember that by increasing the current on a fixed wire diameter, the type of metal transfer across the arc will eventually alter.

What is current density?

This may be described as amps per square mm of the cross-sectional area of the electrode filler wire.

What is the correct open circuit voltage?

This is always the lowest that will give the correct arc voltage. There is usually a 2- to 3-volt drop from open circuit voltage to welding voltage when the arc is struck.

What happens if the open circuit voltage is too low?

This will produce a low arc voltage which will give a very short arc length. This can result in 'stubbing' the electrode into the weld pool, or lack of penetration.

How does too high an open circuit voltage affect the weld?

This causes the arc voltage to be high, resulting in a fierce, long arc length. There will be heavy spatter losses. When using dip transfer, an unsatisfactory sluggish type of metal transfer will result. When welding with spray transfer, the wire electrode may burn back and fuse to the contact tube inside the welding gun nozzle.

Why is wire feed speed important?

Adjustment of the wire feed speed control results in extensive alterations to the welding current. Therefore, by increasing the wire feed speed, the welding current is increased. This in turn allows the electrode filler wire to burn-off at a faster rate, increasing the amount of molten metal passing across the welding arc.

It is of paramount importance that the wire-feed speed be correctly 'tuned' to the voltage setting in order to obtain the correct electrode burn-off rate into the joint.

How is the welding amperage adjusted?

Usually the amperage is preset with the wire feed speed, so that as the wire feed speed control is adjusted, the correct amperage for that feed rate is obtained.

How is the current checked?

During welding the current is indicated by an ammeter in the power source.

What is the effect of current setting too high?

This results in a faster deposition rate, deeper penetration and a large weld bead. The type of metal transfer may well now be of the spray type, and positional welding may prove difficult. Spatter losses will also be heavy.

17

What is the effect of current setting too low?

This results in a smaller weld bead and shallow penetration, and a slower deposition rate. 'Stubbing' of the electrode wire may now occur.

How is the arc length controlled?

With MAGS welding, the arc length is self-adjusting. If the weld passes over a tack weld, the arc momentarily shortens, and the voltage is lowered, allowing the current from the power source to increase. This melts the electrode filler wire at an increased speed until the required arc length is achieved.

What is the correct electrode wire extension?

This is measured from the end of the contact tube to the tip of the electrode wire. For dip-transfer welding the electrode extension

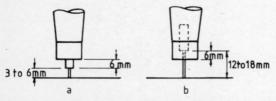

Fig. 9. *Electrode extension for (a) dip transfer and (b) spray transfer*

should be between 3 mm and 6 mm; for spray-transfer welding it should be between 12 mm and 18 mm. See Fig. 9.

Why is electrode wire extension important?

The electrode wire extending from the contact tube will be heated due to its resistance to the flow of welding current. If the

18

extension is too great, the arc current is reduced and shallow penetration results due to the loss in arc power.

What is the correct speed of travel along the weld joint?

The fastest speed possible that will give adequate, neat penetration and correct surface build-up.

What is the result of too fast a weld speed?

This can cause undercut along the top edges, heavy spatter losses and incomplete penetration. The gas shield may be lost slightly, resulting in surface porosity.

What happens if the weld speed is too slow?

When this is with a first run of weld, excessive penetration results; when on a subsequent run and the arc is directed on to liquid metal, there may be a lack of fusion.

What is the recommended welding gun angle?

This should be a slope angle of 70–80°. If this angle is increased to 90°, there will be increased penetration. If the slope angle is decreased, penetration is reduced and the weld surface will be uneven. See Fig. 10.

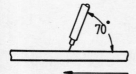

Fig. 10. Slope angle of welding gun

What manipulation may be given to the welding gun?

In general, the deposition of straight beads gives deeper penetration if deposited by the rightward technique (i.e. welding from left to right) because of the efficient use of welding heat. The leftward technique (direction right to left) is preferred for faster weld speeds with reduced penetration. A weaving technique makes wider, less-penetrating welds possible.

Why is the inductance control important?

This improves the weld quality when using dip transfer. Increasing the inductance results in a quieter, spatter-free weld, and the arc is capable of depositing a smooth-surfaced weld deposit. The inductance control should be adjusted to give a minimum of spatter consistent with complete fusion of the weld.

4
MIG, MAG and CO$_2$ welding techniques

What equipment and materials are required to carry out practical exercises?

A MIG welding set connected d.c. electrode positive, welding gun, negative lead, insulated cables, wire snips, wire brush and welding gloves. A welding helmet with the correct shade of filter lens (see BS 679 1982). Reels of electrode filler wire 0.8–2.4 mm diameter. A supply of mild steel plates of 3–6 mm thickness and about 150 mm square.

For correct welding-plant output settings, follow manufacturer's recommendations. Before attempting to weld, a comfortable position must be attained.

Before attempting any welded joints, what practical exercises should be carried out?

On clean plates of 6 mm thickness, practise: (1) arc striking; (2) holding correct slope and tilt angles of torch; (3) moving across the plate at a constant speed; (4) maintaining a correct arc length condition; (5) breaking the arc; (6) wire-brushing after each run; and (7) examination of the deposit.

What preparation is required before arc initiation?

Place a piece of mild steel plate on the welding bench, to which the negative lead is securely fastened. Check the equipment is

21

properly adjusted for dip-transfer welding, and the wire feed speed and voltage set to give a welding amperage of 160 A. The CO_2 gas-flow rate should be adjusted to 12 litres per minute, and the heater switched on.

How should the electrode end be positioned in relation to the contact tip?

The end of the electrode should be cut off to leave a stick-out distance of 5 mm from the contact tube. This is usually carried out with wire snips.

How is the arc initiated?

The electrode is given a slope angle of 70° and a tilt angle of 90° (Fig. 11).

Position the gun with the wire electrode tip 6 mm away from the plate surface, depress the trigger switch and the wire will feed

Fig. 11. *Position of torch when striking the arc*

into the plate surface. The short-circuit effect initiates the arc by allowing the current to flow. When the arc is established, hold the gun steady without any downwards motion. The electrode wire will feed continuously until the trigger switch is released.

22

What is a straight bead?

A bead of weld deposited with the electrode making no sideways movement, the progress of the electrode being only in the direction of welding (Fig. 12).

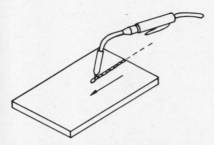

Fig. 12. Depositing a straight bead

What is the procedure when depositing a straight bead?

Using a 6 mm thick clean mild steel plate, strike the arc and hold the nozzle the correct distance away from the plate surface. Remember to hold the correct slope angle with the welding gun. Start by travelling from right to left, without any sideways movement for practising purposes. At the end of the run, break the arc and fill in the crater at the end of the weld bead. Now examine the deposit for correct shape and the presence of any weld defects.

How should the arc be broken at the end of the weld?

To do this, withdraw the arc slowly upwards and backwards to fill in the weld crater. Then release the trigger switch, which discontinues the wire feed, cuts out the current flow and extinguishes the arc.

23

Where are faults most likely to be found?

At the start of a weld, particularly when rejoining to an existing weld bead. These points may show areas of poor fusion and porosity. A lump of crater may be present, which has been caused by wrong slope of electrode, or perhaps by striking too far in front of the existing bead when rejoining.

What is meant by weaving?

This is a side-to-side motion of the electrode across the weld preparation, this being carried out at right angles to the direction of travel (Fig. 13).

Fig. 13. Weaving

When is weaving necessary?

Weaving is used to deposit more metal in one run of weld and also to deposit metal over a wider area.

Are chipping and wire-brushing necessary after each run of weld?

Spatter losses are the main reason for chipping and wire-brushing. A small amount of slag may occasionally be seen on the surface of the deposit, this having been produced from the deoxidizers in the electrode filler wire.

How should a pad of weld be made using straight beads?

Using a piece of mild steel plate of 6 mm thickness and 150 mm square, mark off a 100 mm square in the centre of the plate, indicating this clearly by means of a centre punch.

Deposit a single run of weld along one side of the prepared square, the welding gun having a slope of 70° and a tilt of 90°. The deposit should travel from right to left and be along the edge furthest away from the welder. After brushing and examining, the second run should now be deposited. This should overlap the first by approximately one third, as shown in Fig. 14.

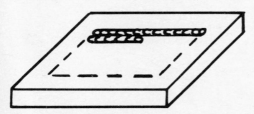

Fig. 14. Pad weld sequence

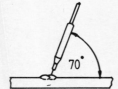

Fig. 15. Tilt angle

When making the second and subsequent runs, the gun should be given a tilt angle of 70° (Fig. 15).

When carrying out this exercise, the pad should be made three layers high, each layer being at right angles to the one below. After cooling, brush the pad and inspect for weld quality and for any defects present.

What further exercises should now be attempted?

Once the above exercises have been mastered, we may now attempt to weld plates together. It is strongly suggested that the following order is adhered to, as progression with each type of

25

weld is desirable. The machine settings are for training purposes only; as skill increases we suggest you alter the welding variables in order to familiarize yourself with the plant settings.

What is a tee fillet weld?

This is a joint as shown in Fig. 16. The plates are usually at right angles to each other.

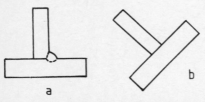

Fig. 16. (a) Fillet weld. (b) Plates tilted ready for welding

Why are fillet welds tilted?

This allows welding to be carried out in the flat position. This means that fewer runs are required, larger-diameter electrode wires may be used, and the weld is completed in a shorter time. All of these factors reduce the overall cost of the work. See Fig. 16.

How should a fillet weld be made in the flat position?

First, two plates 6 mm thick and 150 mm square are cleaned by wire-brushing. They are then tack welded together at each end and positioned for welding. Set the wire feed speed and voltage to give a welding current of 160 A. The electrode filler wire should be 1 mm diameter, and the gas shield CO_2. The gun should travel from right to left, central to the plates being welded, and the slope angle of the welding gun should be 70°. A slight weave of the gun will assist in complete fusion at the edges of the joint. Wire-brush, and then examine the weld before depositing any subsequent runs of weld.

What is the procedure when depositing a weld on an outside corner joint?

Obtain two pieces of mild steel plate, each 150 mm square and 10 mm thick. Clean the plate edges and tack weld at right angles to each other, leaving a gap of 1.5 mm between the plates.

The edges which the root run is deposited upon are comparatively thin compared with the rest of the joint; therefore a slight reduction in the wire feed speed and welding voltage is necessary to reduce the risk of burn-through. The electrode filler wire should be 1.2 mm diameter. The wire feed speed and voltage should be regulated to give a welding current of 140 A.

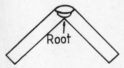

Fig. 17. *Outside corner joint*

Deposit the penetration bead. Weaving is not necessary on the first run. For any subsequent runs of weld, the wire feed speed and voltage should be increased to give around 160 A. A definite weave should not be given. Clean off spatter, wire-brush and inspect. See Fig. 17.

What is a single vee butt joint?

This is shown in Fig. 18.

Fig. 18. *Single vee joint, ready for welding*

Which is the most difficult run of weld to deposit?

This is the root penetration weld bead. Care must be taken with plant settings, correct angles of gun, arc length and speed of travel along the joint. Weaving is not recommended with this run of weld. Consistently good penetration is vitally important, and this run of weld requires much practice. With a 1 mm diameter wire electrode, set the wire feed speed and voltage to give a welding current of 160 A. Deposit the root run.

How should this run appear?

Reasonably flat without any lumps or craters. The edges of the weld should be fused into the sides of each plate. Penetration

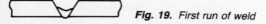

Fig. 19. First run of weld

should be fused through to a depth of around 1.5 mm on the underside of the joint. See Fig. 19.

How should the second run of weld be deposited?

Increase the wire feed speed and voltage to give a welding current of 170 A. Care must be taken to ensure complete penetration and inter-run fusion at the top of each plate, and this is assisted by using a definite weaving technique with the welding gun. The weld surface should be slightly convex and have a definite reinforcement of 10 per cent of the plate thickness.

What procedure is used to weld a butt joint between two pieces of 3 mm mild steel plate?

An open square butt joint should be assembled leaving a gap of 2 mm. Clean the plates and tack weld every 100 mm along the joint. Using a 1 mm diameter wire electrode, select the wire feed speed and voltage to give a welding current of 110 A. Strike the arc at the end of the joint and deposit the weld. The contact-tube-to-plate distance should be 5 mm, and the speed of travel adjusted to give good penetration and a uniform weld surface.

What is a horizontal-vertical fillet weld?

This is shown in Fig. 20. Note that the base plate is flat on the bench and the plate to be welded to it is vertically positioned at 90°

Fig. 20. Horizontal-vertical fillet weld

How should a horizontal-vertical fillet weld be deposited?

Two mild steel plates, each 150 × 75 × 6 mm, are used to practise this weld. The plates are wire-brushed clean and positioned; a tack weld at each end will hold them in alignment. Using a 1.2 mm diameter electrode wire, set the wire feed speed and voltage to give a welding current of around 180 A. The arc may then be initiated. The welding gun slope angle is again 70°, but

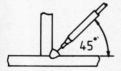

Fig. 21. Tilt of gun and shape of first run of weld

notice that the tilt angle is now at 45° to the base plate. Maintain these angles continuously while depositing the root run of weld. Weaving is not necessary. Chip away spatter and wire-brush. See Fig. 21.

29

If more than one run is required, how should subsequent runs be deposited?

After the first run, slightly increase wire feed speed and amperage settings. Deposit the second run of weld between the base plate and the first run, overlapping the latter by about one third. The electrode tilt angle should be altered to 60°, the slope angle remaining at 70° (Fig. 22). Weaving is not required.

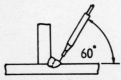

Fig. 22. *Second run of weld; note the tilt of the electrode*

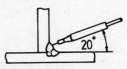

Fig. 23. *Third run of weld; note further change of tilt of the electrode*

Wire-brush the second run of weld and deposit the third run between the vertical plate and the second run of weld. The tilt angle for this run should be around 20° (Fig. 23). All that now remains is to wire-brush and examine for weld quality and defects.

What is a lap joint?

The plates are fitted as shown in Fig. 24. It can be welded in the same way as the horizontal-vertical fillet weld.

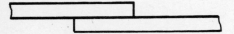

Fig. 24. *Lap joint*

What step should be taken after completion of these exercises?

All the welds practised up to now have been done by dip transfer. We now suggest that the exercises are repeated using spray transfer. Remember that the wire feed speed, the amperage and the voltage will be increased, so you should consult manufacturer's recommendations for these settings. Since more metal will be deposited, and the depth of penetration will be greater, an increased weld speed is necessary.

LAP AND FILLET WELDS IN THE VERTICAL POSITION

How should a vertical fillet weld be deposited?

Use two pieces of mild steel, each 150 mm × 150 × 10 mm. Set up the fillet joint placing a tack weld at each end. Using a dip transfer technique with a welding current of 130 A and an electrode wire diameter of 1 mm, the root run of weld is deposited by travelling in a vertical downwards direction (over 10 mm thickness, the root run should travel in a vertical upwards direction). It is important to achieve good penetration and fusion along the plate faces. A slight weaving action may be used to keep control of the weld pool. See Fig. 25.

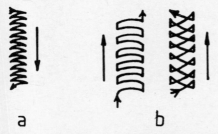

Fig. 25. (a) Weave for root run (b) Subsequent weave

Wire-brush and examine the surface of the root run. Increase the wire feed speed and voltage to give a welding current of 130 A. Then deposit the second layer by welding vertically upwards. A distinct weave, pausing at the edges of the joint but passing rapidly across the weld face, is desirable (see Fig. 25). Remove spatter, wire-brush and examine the completed weld.

31

Are any alterations in technique required when lap joints are welded vertically upwards?

Basically, the same technique should be employed. Extreme care is necessary to avoid burning the edge of the top plate, which would reduce the mechanical properties of the joint.

HORIZONTAL-VERTICAL BUTT JOINT WELDING

When is this weld position used?

It is used where it is not possible to turn the fabrication over to allow welds to be made in the flat position. It is regularly used on site work.

What is the horizontal-vertical welding position?

The weld is deposited parallel to the horizontal. See Fig. 26.

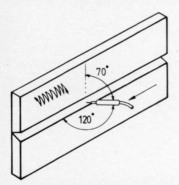

Fig. 26. *Horizontal butt joint welding, showing weaving action*

How should a butt joint be carried out in this position?

The edges of two 10 mm thick plates are prepared to form a 60° single vee preparation. The plates are then tack welded together leaving a root gap of 2 mm. Using a 1 mm diameter wire electrode, with the wire feed speed and voltage adjusted to give 140 A, deposit the root run. A slight backwards-and-forwards action helps control penetration. The tilt angle of the torch for the root run should be 70°, the slope angle being 120°.

After brushing, the second run is deposited between the root run and lower plate. a third run of weld is now deposited between the second run of weld and the top plate. With these subsequent runs a weaving action is generally used.

What weaving techniques are recommended?

These can be seen in Fig. 26.

OVERHEAD BUTT WELDING

What preparation is required when welding 6 mm mild steel plate?

A single vee butt joint with an included angle of 60° and having a root face and root gap of 1.6 mm. Plate edges must be cleaned and free from grease.

How is such a joint welded?

Select welding conditions to give dip transfer. The wire diameter should be 1 mm, and the wire feed speed and voltage adjusted to give a welding amperage of 120. The root run should be deposited without a weaving action, the speed being adjusted to

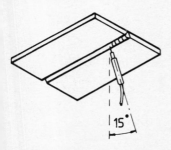

Fig. 27. *Overhead butt weld*

33

give good fusion at the joint faces. Any irregularities of the root run should be removed by lightly grinding before depositing any subsequent weld runs. Subsequent runs should then be deposited using a circular weaving technique to ensure complete fusion at the weld edges. Finally, wire-brush and examine. (See Fig. 27).

OVERHEAD FILLET WELDING

What is the procedure for carrying out a fillet weld in the overhead position between two pieces of 6 mm mild steel plate?

The joint should be tack welded at each end and placed in position. Select welding conditions for dip transfer. Wire diameter should be 0.8 mm, and wire feed speed and voltage should

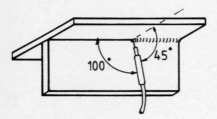

Fig. 28. *Overhead fillet weld*

be adjusted to give a welding current of 120 A. Deposit a three-run fillet weld without weaving. Adjust the welding gun angle of tilt for each run of weld (this is similar to depositing the same type of joint in the horizontal-vertical position). See Fig. 28.

WELDING OF PIPELINES

What are the limitations of MAGS welding on pipelines?

MAGS welding is generally used on mild steel pipelines in excess of 100 mm diameter and having a wall thickness of not less than 4.5 mm.

All root runs and pipes welded in a fixed position should be welded using dip transfer.

What joint preparation is recommended?

A single vee with a 60° vee included angle, 1.5 mm root face and a root gap which is dependent on the pipe wall thickness. Cleaning and degreasing is advisable before welding.

What are the recommended plant settings?

The wire extension from the contact tube may be increased to 10 mm to facilitate access. Using a 0.8 mm diameter electrode wire, adjust the wire feed speed and voltage control to give a welding current of 110 A. This will provide dip transfer conditions.

What type of electrode wire should be used?

This should comply to BS 2901, namely it should be 0.8 mm diameter and contain the deoxidizing agents silicon and manganese.

Is preheating necessary?

For pipe walls below 12 mm thickness a preheat is not necessary.
 Between 12 and 18 mm wall thickness, a low preheat of 150°C will control stresses and distortion tendencies.
 For over 18 mm wall thickness, the preheat may be increased to 200°C.

How are pipes aligned before welding?

By tack welding in four places at 90° intervals. The tack welds should be at least 12 mm long, and be fused to the full depth of the pipe wall. Tack welds are often shaped to allow easy weld pick-up (Fig. 29).

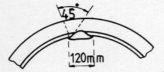

Fig. 29. Tack weld shape

Sometimes internal alignment clamps are used; these eliminate the need for tack welding.

Can rollers be used when welding pipes?

Yes, if the work is small enough to be positioned on rollers. The pipes are rotated at a regular speed, allowing welding to be carried out in the flat position at one spot only. The speed of rotation should be equal to the normal welding speed.

How is the weld made when the pipes are in the fixed horizontal position?

To deposit the root run, strike the arc at the 12 o'clock position and proceed to weld downwards. The slope angle of the welding gun should be 70° to a tangent to the weld pool (Fig. 30). It is essential to maintain the correct angle, since too steep an angle causes excess penetration whilst too shallow an angle creates a deflection to the shielding gas flow, allowing porosity to occur.

As the weld progresses downwards, a slight side-to-side weave may be used to give good fusion. Upon reaching the inclined position, the slope angle of the welding gun is increased to 75°, which is maintained until the 6 o'clock position is reached.

If welding is stopped, this should be at a tack weld to minimise stresses. Before beginning the run of weld on the opposite side of the pipe, the start and the finish of the first run should be tapered to remove any crack or unfused metal present. Restrike the arc at the 12 o'clock position, about 12 mm in front of the first start, and quickly work back into the first deposit and then proceed downwards in the opposite direction. The weld should then be wire-brushed and inspected.

36

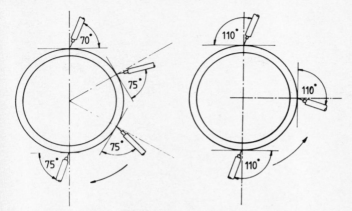

Fig. 30. *Slope angle of welding gun for deposition of root run*

Fig. 31. *Slope angle of welding gun for second run, using vertical upwards method*

The second run of weld should be deposited using the vertical upwards method. Strike the arc at 6 o'clock, the slope angle being 110° in the direction of travel (Fig. 31). When the 12 o'clock position is reached, the arc is extinguished. The start and finish points of this run, respectively the 6 and 12 o'clock positions, should now be shaped to allow smooth pick-up. Wire-brushing and removal of any spatter losses completes the job.

5
Weld defects when semi-automatic welding

What weld defects may be found in a joint?

Lack of penetration, porosity, incomplete fusion, undercut, spatter losses and cracking.

What are the causes of insufficient penetration?

Insufficient penetration (Fig. 32) may be caused by: (1) The electrode being wrongly directed in relation to the centre of the weld preparation. This can be avoided by concentrating the arc heat to the centre of the joint. (2) The joint may not have been

 *Fig. 32.* Lack of penetration

prepared properly, e.g. the root face is too thick, the root gap is too narrow or the included angle of bevel is insufficient. (3) Incorrect slope angle of the welding gun. This angle may need to be increased or decreased during actual welding to control penetration (Fig. 33). (4) Incorrect speed of travel along the joint. Too fast a weld speed means there will not be sufficient time for the weld preparation to melt, resulting in poor penetration. If the weld speed is too slow, excessive penetration will occur. (5) A badly worn contact-tube tip. This causes a bad arcing condition which hinders penetration. (6) Too low a welding amperage or

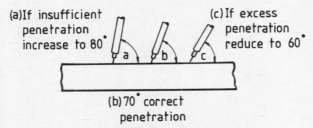

(a) If insufficient penetration increase to 80°

(c) If excess penetration reduce to 60°

(b) 70° correct penetration

Fig. 33. Effect of slope angle on penetration

too large an electrode wire diameter. This results in a sluggish type of arc which does not allow penetration. (7) Irregular wire feed speed. This can be solved by re-adjustment. (8) An excess wire stick-out from the contact tube tip. This causes an increase in the circuit resistance, reducing the amperage and therefore the penetration.

What is undercut?

This is a reduction of the cross section of the plates being welded (See Fig. 34).

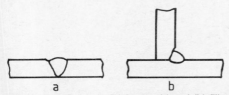

a   b

Fig. 34. Undercut in (a) butt weld and (b) fillet weld

What are the causes of undercut?

Undercut may be caused by the wire feed speed and amperage control being set too high, the welding speed being too fast, or by poor manipulation of the welding gun.

A small amount of intermittent undercut may be acceptable if it is not too deep. It can be repaired by depositing a further layer of weld over the undercut.

What is porosity?

This can be seen in Fig. 35.

 **Fig. 35.** Porosity

How may porosity have been caused?

Porosity may be the result of: (1) The base plates not being cleaned of paint, oil, grease, rust and scale. Arc heat can cause the above impurities to gasify and become trapped in the weld metal during solidification. (2) The electrode filler wire being wet or perhaps rusty. This should be stored correctly before use. (3) The shielding gas flow rate being either too high or too low. Blockage of the welding gun nozzle is one possible cause of too low a gas flow rate. (4) The nozzle-to-work distance being too long, resulting in an excess electrode stick-out. This causes the gaseous shield to operate over a wider area, thereby reducing the gas density. Atmospheric gases may now penetrate into the weld pool, resulting in porosity. (5) Loose gas connections or a split gas hose which allow air to mix with the shielding gas passing through the hose and to the weld pool. (6) The use of a non-syphon type CO_2 cylinder. This will result in increasing moisture in the gas shield as the cylinder empties. (7) An inefficient or malfunctioning heater resulting in loss of shielding gas due to the regulator icing up.

What causes poor fusion?

This is the result of incomplete melting-together of the edges being welded and the filler metal being deposited (Fig. 36). This

40

Fig. 36. Poor fusion

may be because of the voltage, wire feed speed or amperage settings being too low, or use of the welding gun at the wrong slope angle.

What are the causes of spatter losses?

Spatter losses may be the result of welding on rusty plate surfaces, having too high or too low a voltage setting or by incorrect use of the inductance control. A wrong type of shielding gas when welding mild steel with spray transfer results in heavy spatter losses.

How may spatter be controlled?

Spatter may be controlled by cleaning the plate surfaces before welding, adjusting the correct, recommended voltage setting and making use of the inductance control. Spatter encrustation of the nozzle can be reduced by spraying the nozzle with an anti-spatter silicone oil before welding.

Why do welds sometimes crack on cooling, and how can this be avoided?

1. Insufficient metal deposited in the weld crater. This can result in cracks which may well spread through any further runs of weld. To avoid this, fill in the weld crater upon completing the weld.
2. If the weld is deep and narrow, contraction stresses will be severe which may lead to a crack occurring. This is particularly so if the plates being welded are restrained.
3. If there is insufficient metal deposited in the root run to withstand contraction stresses during cooling of the weld, cracking will occur. Avoid by depositing sufficient weld metal.

41

4. Using spray transfer with too high a welding current may result in cracking.
5. The final shape of the weld should ideally be slightly convex; if it is concave, cracking may occur.
6. If low-alloy high-tensile steels are being welded, a rapid rate of cooling will increase the possibility of a crack occurring. This may be avoided by applying a preheat of 200°C.
7. The electrode composition must be compatible with, and have similar mechanical properties to, the base metal.

6
MIG welding of various materials

ALUMINIUM AND ITS ALLOYS

What are the general characteristics of aluminium?

It is light in weight, having a specific gravity of only 2.6 (mild steel is 7.8). It has very good thermal and electrical conductivity and offers very good resistance to corrosion.

What are the welding difficulties associated with aluminium?

It has a low melting point of 658°C. There is no colour change when aluminium is heated to its welding temperature; this can make it difficult during welding to judge when melting is about to take place.

It has a high thermal conduction rate. This can be overcome by providing a high heat input during welding. Indeed, heavy sections may well require the assistance of a preheat before welding.

A tenacious oxide film forms on the surface of the aluminium. This is of a refractory nature and is difficult to remove by welding heat due to the oxide having a much higher melting point than the aluminium. In fact, as the temperature increases, the thickness of the oxide film also increases. It must, however, be removed before welding can successfully take place.

How is the oxide film removed?

This may be removed by pickling in an acid solution such as a 10 per cent phosphoric acid solution at a temperature of 30°C, and by wire-brushing or filing before welding by any process. When manual metal arc or oxy-acetylene welding, use a flux to dissolve the oxide film (alumina). But when MIG or TIG welding aluminium, a cathodic cleaning action within the arc removes the oxide film. See page 78.

When should the MIG or TIG welding processes be used?

Metal thickness is the determinant. Below 3 mm thickness TIG welding is the recommended process. Above this thickness, MIG welding should be used.

What are the advantages of welding aluminium by the MIG welding process?

The main advantages are:

1. It is an all-positional welding process.
2. There is a small heat-affected zone.
3. Welds are of excellent appearance and of high quality.
4. The deposition rate is extremely fast.
5. Welds are made without the need for a corrosive flux.

How should the joint be prepared for welding?

Cleanliness is essential, so the parts to be welded should be degreased and the surface oxide removed. If welding is not carried out immediately, the cleaned edge surfaces should be covered with a cellulose adhesive tape, in order to prevent the oxide reforming before welding is carried out.

What edge preparations are recommended?

For Butt welds from 3 mm to 5 mm thickness, use a closed square edge butt joint, leaving a gap of half the plate thickness (Fig. 37a).

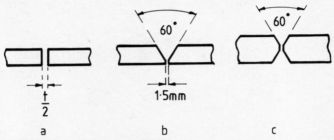

Fig. 37. (a) Square edge butt joint. (b) Single vee butt joint. (c) Double vee butt joint

From 6 mm to 15 mm, a 60° included angle single vee preparation with a root gap of 1.5 mm should be used (Fig. 37b).

Above 15 mm consider a 60° double vee preparation (Fig. 37b).

Should the underside of the weld have any backing?

Yes, because higher production is attained by use of a temporary backing bar. These are usually made from stainless steel and are

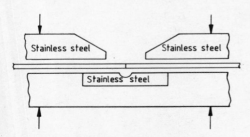

Fig. 38. Backing arrangement

45

positioned on the underside of the joint (Fig. 38). Higher welding speeds and use of higher amperage settings are now possible without fear of excessive burn-through.

What gas shield is necessary?

This should be pure argon.

What type of welding gun is preferred?

This should be the reel-on gun type in order to avoid 'kinking' of the soft aluminium wire. The gun may be either air- or water-cooled.

What precautions should be taken with the electrode filler wire?

Before packing this should have been 'shaved' to remove any impurities picked up during manufacture. Before use the wire should be stored in air-tight bags; during welding, spool covers should be kept over the wire.

What is the composition of the electrode filler wire?

This is indicated in BS 2901, part 4.
 For welding pure aluminium use GI pure aluminium wires.
 Aluminium-magnesium alloys should be welded with NG61 aluminium/5 per cent magnesium alloy.
 Aluminium-silicon alloys and all aluminium castings should be welded with NG21 aluminium/5 per cent silicon alloy.

What type of metal transfer is used?

This will be spray, or pulsed.

Where should the arc be struck?

Cleaner starts are attained by using a run-on plate, and weld craters at the end of the weld are avoided when a run-off plate is

Strike

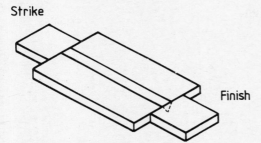

Finish

Fig. 39. *Run-on plate and run-off plate*

used. If these are not available, strike the arc 25 mm from the start of the weld joint and travel backwards to the start of the weld. This provides a preheat effect. See Fig. 39.

How is the arc struck?

For aluminium, the switch is depressed and argon gas begins to flow; this purges the weld area. The electrode wire is then scratched on to the surface of the work to be welded. This action strikes the arc, and the electrode filler wire automatically begins to feed.

COPPER

What are the different types of copper?

These are tough pitch copper and deoxidized copper.

47

What is tough pitch copper?

This is copper which contains 0.5 per cent of oxygen. It is not recommended for welding due to its tendency to crack during welding.

What is deoxidized copper?

The oxygen has been removed during manufacture by the use of deoxidizing agents such as phosphorus and silicon. This type of copper can be fusion-welded.

What are the difficulties associated with the welding of copper?

These include 'hot shortness', high conduction rate, high expansion rate and rapid oxidation.

What is 'hot shortness'?

Certain metals when heated to just below their melting point suffer a considerable loss of ductility, and thus become brittle. Thus at this temperature the weldability of such a metal may well be limited, and if it is stressed (e.g. by forces of expansion and contraction), cracking will result.

How are the high conduction and high expansion rates controlled?

These are the main problems when welding copper. Above 4 mm thickness a suitable preheat will assist in overcoming the problem.

The amount of preheat required depends on the thickness of the metal being welded. For example:

4 mm to 6 mm thickness: preheat to 125°C.
6 mm to 10 mm thickness: preheat to 175°C.

10 mm to 15 mm thickness: preheat to 375°C.
Above 15 mm thickness: preheat to 475°C.

It is important that the preheat temperature is maintained while welding is carried out.

What gas shield is used?

Nitrogen may be used when welding copper in the flat position. Good results are obtained by using argon/nitrogen gas mixtures.

For high-quality welds made in the flat position or any other position, argon is the preferred gas shield.

What joint preparation is required when welding copper?

Before welding, all oil and grease should be removed; scratch wire-brushing is necessary both before welding and also before depositing any subsequent runs of weld.

Up to 4 mm thickness, use an open-square butt joint leaving a 1.5 mm gap between the plates (see Fig. 40a).

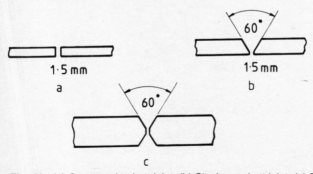

Fig. 40. (a) Square edge butt joint. (b) Single vee butt joint. (c) Double vee butt joint

4 mm to 18 mm thickness, use a 60° single vee leaving a 1.5 mm root face and gap (Fig. 40*b*).

Above 18 mm thickness, use a 60° double vee leaving a 1.5 mm root face and gap (Fig. 40*c*).

What weld backing should be used?

Support must be provided for butt joints during welding, and this is achieved by using mild steel backing bars. These are tightly clamped on the underside of the joint supporting the weld, and help to avoid distortion.

What type of electrode filler wire is used?

This should comply to BS 2901 part 3. The electrode wire should contain adequate deoxidizing elements; a grade C7 filler wire electrode is recommended.

How should the welding gun be adjusted?

The contact tube should be recessed 10 mm inside the nozzle and the electrode filler wire should protrude 25 mm from the contact tube.

How is the arc initiated?

The arc initiation switch on the control box unit is positioned to give a scratch start. This means that when the welding gun trigger switch is depressed, the shielding gas flows and the current contactor closes but the electrode wire does not feed until the electrode tip is contacted to the work. This method of arc initiation is normally used when the materials being welded are good electrical conductors.

What type of metal transfer is used?

Welds made with a nitrogen gas shield are acceptable when MIG welding copper in the flat position. When using nitrogen alone, the metal transfer across the arc is of a coarse globular type. Improved results are obtained with an argon/nitrogen gas mixture. For positional welding, which requires a smooth type of spray transfer, argon alone is recommended.

Which technique of welding is used?

The backhand or rightward technique should be used. The slope and tilt angles of the welding gun are shown in Fig. 41.

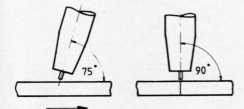

Fig. 41. *Slope and tilt angles when MIG welding copper*

STAINLESS STEEL

What varieties of stainless steel are available?

These are ferritic, martensitic and austenitic stainless steel.

What is ferritic stainless steel?

This contains 16 to 30 per cent chromium, a maximum of 0.1 per cent of carbon and an iron base. The high chromium content ensures a high resistance to corrosion. This alloy is not heat-treatable. However, if welding is attempted, grain growth will occur in the heat-affected zone. If welding is necessary, a preheat temperature of 150°C is recommended. The filler material composition should ideally be 25 per cent chromium and 20 per cent nickel. After welding, the work should be post-heated to 750°C, followed by slow cooling.

What are the uses of ferritic stainless steels?

Because of the pronounced increase in corrosion-resistance of these alloys, they are widely used in the chemical industry, and for domestic purposes such as stainless steel sinks, food containers, refrigerator parts and cutlery.

What is martensitic stainless steel?

This contains 10 to 14 per cent chromium and up to 0.2 per cent carbon. It is an air-hardening steel, which means that even with relatively slow cooling rates in air, the hard, brittle martensite will form. This structure may well create problems of cracking during welding. If welding has to be carried out, the procedure should be similar to when welding the ferritic variety.

What are the uses of martensitic stainless steel?

It is used in situations where high hardness and corrosion-resistance is important, e.g. surgeons' knives and dental instruments.

What is austenitic stainless steel?

This contains up to 30 per cent chromium, 25 per cent nickel and below 0.25 per cent carbon. It may also be regarded as a heat-resisting steel. The presence of nickel in excess of 6 per cent will prevent any change from the austenite condition. Regardless of the speed of cooling, the material will not be hard and brittle.

What are the uses of austenitic stainless steel?

Austenitic stainless steel is readily welded, and therefore is used extensively in the manufacture of chemical and nuclear plants, and welded beer barrels.

How can the weldable variety of stainless steel be identified?

By means of a magnet. Ferritic and martensitic stainless steel are magnetic, whilst austenitic stainless steel, the weldable type, is non-magnetic.

What properties affect the welding of austenitic stainless steel?

The material melts around 1420°C. If the metal is not protected during welding, rapid oxidation will occur at this temperature.

The material has a high expansion rate along with a low conduction rate, so extra care must be taken to avoid distortion.

Carbon pick-up from any source must be avoided, otherwise intergranular corrosion may result.

With these factors well understood, the material is readily welded.

What is intergranular corrosion?

When austenitic stainless steel is heated between 400°C and 800°C, the carbon combines with chromium forming chrome carbides. These are insoluble in the austenitic solid solution, and are rejected to the grain boundaries thus leaving the areas adjacent to the boundaries deficient in chromium. This results in

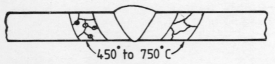

Fig. 42. *Intergranular corrosion*

the iron within these areas being oxidized, allowing corrosion to take place. This problem does not occur in the weld metal itself but in the heat-affected zone adjacent to the weld, and thus is often referred to as weld decay. See Fig. 42.

How is the problem avoided?

The entire job after welding could be reheated to 1000°C, thus taking the chromium back in solution, and then followed by quenching. Obviously the size of the work imposes limitations on this method.

The carbon content could be reduced during manufacture, but this would increase the purchase price.

The method widely used industrially is the addition of stabilizing elements during manufacture. The stabilizers added may be either titanium or niobium. Nowadays most stainless steels are sold in the stabilized condition.

What preparation is necessary before welding?

The edges to be welded should have all oil, grease and dirt removed. To provide good penetration and easy access for the welding gun, the following is recommended:

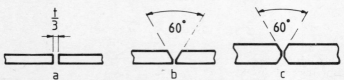

Fig. 43. *Preparation for thickness (a) up to 3 mm, (b) 3 to 10 mm, and (c) over 10 mm*

54

Up to 3 mm thickness, use an open-square butt joint, the gap being one third of the metal thickness (Fig. 43a).

From 3 to 10 mm thickness, use a 60° single vee with a 1.5 mm root face and gap (Fig. 43b).

Above 10 mm thickness, use a 60° double vee with a 1.5 mm root face and gap (Fig. 43c).

Is backing necessary?

Backing with suitably grooved copper bars ensures neat, sound welds. To prevent oxidation of the underside of the weld, argon gas may be fed along the backing groove (Fig. 44).

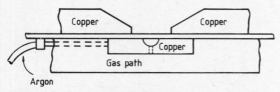

Fig. 44. *Backing complete with argon supply (BS 3019)*

What electrode wires are used?

These are given in BS 2901 part 2.

The corrosion-resistant steel wires are A8Nb. This is a niobium-stabilized, chrome-nickel composition.

What gas shield is required?

This should be argon plus 1 per cent oxygen (see page 64). Gas mixtures containing CO_2 must not be used, as intergranular corrosion may result.

Is tacking ever used?

If jigs cannot be used, close tack welding is essential, as this assists in keeping correct alignment. The tack welds should be longer than for similar thicknesses of mild steel.

What type of transfer is used?

Good results are obtained wth spray transfer. A controlled (pulsed) spray transfer may also be used; this is referred to in Chapter 1.

How is the arc struck?

The arc initiation on the control unit is positioned to provide a feed start. The torch nozzle is held approximately 10 mm from the work and the trigger switch on the gun is depressed. The gas then flows (pre-purge), then the wire feeds, and as the wire contacts the surface of the work, the arc is established.

Which technique is used?

The forehand or leftward technique is preferable when welding in the flat position. Weld craters should be filled at the end of the weld to reduce the risk of cracking.

NICKEL

What are the commonly used nickel alloys?

These are:

Grade	Constituents
Nickel 200	Pure nickel
Monel	Nickel-copper alloy
Inconel	Nickel-chromium-ferrite alloy
Nimonic	Nickel-chromium-titanium alloy

How should the edges be cleaned before welding these alloys?

All burrs must be removed from cut edges, which should then be cleaned by wire-brushing an area of 25 mm along each edge to be welded. This is followed by degreasing.

What type of joint preparation is recommended?

For thicknesses of 1.6 mm to 3 mm, use an open-square butt joint, leaving a gap of 1.6 mm between the edges. For 3 mm to 6 mm thickness, a 70° single vee butt joint with a 1.6 mm to 3 mm root gap and a 1 mm root face is recommended. This should be followed by back-chipping and the deposition of a sealing run of weld.

Why must sulphur pick-up be avoided?

Sulphur pick-up invariably results in the completed welded joint cracking. Before welding, cleaning with spirits assists in controlling this problem.

Why is back-chipping recommended?

This removes any unfused areas along with any penetration defects from the underside of the weld. The deposition of a sealing run will ensure sound weld metal throughout the deposit.

Is weld backing advisable?

Jigging and backing, as recommended in BS 3019, controls distortion and reduces penetration defects.

Is preheating necessary?

Thin sections of nickel and its alloys are welded without a preheat. Thick sections are assisted by a low preheat not exceeding 150°C.

What gas shield is recommended?

This should be argon. Argon-helium mixtures are widely used in the USA; these tend to deposit a wider, flatter weld bead.

What gas flow rate is recommended?

The ideal welding conditions are attained at around 25–30 litres per minute.

What type of metal transfer is used?

Below 6 mm thickness, dip transfer is advised. Above this, spray transfer provides the ideal conditions.

How is the arc struck?

This again is a feed start, and is similar to the method used to initiate the arc when welding stainless steel (see page 56).

What electrode wires are used?

These should comply to BS 2901 part 5.
  When welding nickel, it is usual to use an electrode filler wire containing a small percentage of titanium. This helps to avoid porosity in the completed weld.

7
Gas shields for MIG, TIG and CO_2 welding

What is the purpose of the gas shield?

The gas shield has several functions, including protecting the weld metal from the atmosphere during welding, assisting with penetration, influencing the weld surface shape, and providing arc stability by ionizing the arc gap.

What are the shielding gases available?

These include argon, helium, nitrogen and carbon dioxide. Argon-rich mixtures such as 95 per cent argon + 5 per cent CO_2, 80 per cent argon + 20 per cent CO_2 and argon + 1 or 2 per cent oxygen are used on certain ferrous alloys. Others may be obtained to customers' requirements.

When is argon used for shielding?

Argon is an inert gas and is used for shielding purposes for both the MIG and TIG welding processes.

When MIG welding, argon is used successfully on non-ferrous materials such as aluminium. The electrode filler wire is connected d.c. positive, and a good cleaning action is obtained along with favourable metal transfer conditions.

Argon gas may also be used when TIG welding for both ferrous and non-ferrous materials.

What is an inert gas?

An inert gas does not combine chemically with any other known element. It does not burn, nor support combustion.

Inert gases used for welding are argon and helium. Other inert gases are neon, crypton, xenon and radon; these are not used when welding because of poor availability and high costs.

What advantages does argon have as a shielding gas?

Argon does not combine with the metal being welded, it is insoluble in weld metal and offers only a slight resistance to current flow. This results in a smooth, soft stable arc condition with a relatively low arc voltage compared with other gases. Good, clean welds are possible on aluminium, magnesium, copper, nickel and steels.

When is helium used?

It may be used when both MIG and TIG welding. It offers a higher resistance to current flow than argon, resulting in a greater voltage drop across the arc, which in turn increases the depth of

a   b

Fig. 45. Weld resulting from use of (a) argon and (b) helium. Note the deeper penetration resulting from use of helium

penetration (Fig. 45). However, a drawback to the use of helium in Great Britain is its high cost. It is used frequently in the USA, where it is found as a natural gas and is cheaper to use than argon.

Has nitrogen any use with these processes?

Nitrogen combines with a number of metals forming hard, brittle compounds. It does not, however, combine with copper at welding temperatures, and may be used as a shielding gas when MIG or TIG welding this metal.

When MIG welding copper, only globular transfer is possible. This results in the finished weld surface appearing somewhat rougher than when copper is welded using argon as a gas shield. However, this rough surface does not interfere with weld strength as the penetration should be acceptable. It is worth remembering that nitrogen is only a fraction of the purchase price of argon. Success has been achieved with argon-nitrogen mixtures which are economical.

What use is made of carbon dioxide (CO_2) as a shielding gas?

This is suitable when MAG welding mild steel. It is used with dip transfer, and a good short-circuiting arc condition is obtained.

What is MAG welding?

This is welding with metal-active gas. The gas shield dissociates at welding arc temperatures.

How does CO_2 react within the arc during welding?

CO_2 is an active gas and at welding arc temperatures it decomposes to form carbon monoxide + oxygen, which provide an oxidizing arc atmosphere. The electrode filler wire should contain an excess of deoxidizing elements such as silicon, manganese and aluminium. These combine with the released oxygen, forming a

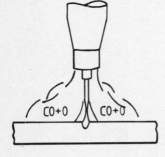

Fig. 46. *Decomposition of CO_2*

thin glass-like slag on the weld surface, but the weld metal is not oxidized. See Fig. 46.

How pure must CO_2 be for welding?

It must be a minimum of 99.9 per cent pure. The moisture content of the CO_2 gas must not exceed 150 parts per million, otherwise weld porosity will result. If as much as 1 per cent nitrogen is present, again porosity will occur.

What is the maximum discharge rate from a CO_2 cylinder?

The discharge rate must not exceed 25 litres/minute. If a higher rate is necessary, a manifold system of cylinders must be used.

Why is this?

Above 25 litres/minute, the liquid CO_2 in the cylinder cannot vaporize quickly enough; this will freeze up the regulator even if a 150 W heater is used.

Why is CO₂ not recommended for spray transfer?

The higher currents required for spray transfer with a CO_2 shield would result in a fierce, fiery arc condition with heavy spatter losses. This is because of a non-axial type of transfer.

What gas mixture is advisable for spray-transfer welds on mild steel?

Good spray-transfer welds are made using an argon-rich gas mixture such as 95 per cent argon/5 per cent CO_2, or 80 per cent argon/20 per cent CO_2. With these mixtures the magnetic pinch effect on the wire electrode is more pronounced than with just CO_2 alone. This now allows spray transfer to be used at lower current densities, which results in an axial type of transfer giving reduced spatter losses. See Fig. 47.

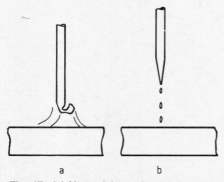

a   b

Fig. 47. *(a) Non-axial transfer. (b) Axial transfer*

An axial form of metal transfer occurs when MAGS welding with a shielding gas that does not decompose at arc heat temperatures. This allows the arc column to remain concentric to the electrode wire.

What is non-axial transfer?

This occurs when MAGS welding using a shielding gas that disassociates at arc heat temperatures. An example of such a gas is CO_2 (see page 61).

When are argon-oxygen mixtures used?

When MIG welding stainless steel, argon/1 per cent oxygen is most successful. The small addition of oxygen assists the flow of weld metal, promoting favourable transfer with a minimum of spatter loss. The oxygen addition 'washes', or reduces the surface tension of, the weld pool. This results in an even flow of metal along the fusion edges, providing a neat undercut-free weld.

The 1 per cent oxygen in the gas mixture causes the heated surfaces of the joint preparation to oxidize slightly, which is then counteracted by the deoxidizers in the electrode filler wire. See Fig. 48.

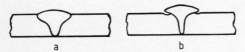

a   b

Fig. 48. *Weld using (a) argon–oxygen mixture; (b) pure argon*

Argon/2 per cent oxygen mixtures are recommended when dip transfer is being used for stainless steel, providing there are sufficient deoxidizers present in the electrode filler wire.

Are argon-oxygen mixtures used when TIG welding?

No, because the oxygen addition would oxidize the tungsten electrode during welding.

Are any other gas mixtures available?

Yes. Argon-hydrogen mixtures are sometimes used when TIG welding heavy stainless steel or nickel fabrications.

Is a flux sometimes used when CO$_2$ welding?

Yes; flux-cored electrode wires are available.

When would a flux-cored wire be used?

It has been mentioned that spray transfer with a CO$_2$ shield gives a fierce arc with heavy spatter losses. This can be overcome by

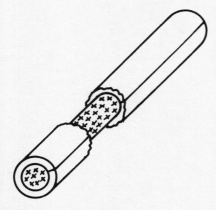

Fig. 49. *Flux-cored wire*

using a flux-cored steel electrode filler wire, along with the CO$_2$ gas shield (Fig. 49).

What advantages are obtained from flux-cored wires?

These are numerous and include: (1) reduction in spatter with improved weld profile; (2) efficient interfusion along the toes of the weld; (3) improved metal-transfer conditions; (4) fast deposition rates with higher welding currents; (5) high metal-recovery rates; (6) lower cost than argon-rich mixtures on steels.

What does the flux core contain?

Deoxidizers, slag-forming materials, arc stabilizers and additional alloying elements may be added. All of these will improve the mechanical properties of the weld material.

8
Recommended joint preparations and weld symbols

Why should joints be prepared before welding?

Correct joint preparation ensures that adequate penetration and fusion are possible, and that the right amount of weld metal is deposited to ensure correct strength and reinforcement of the welded joint.

Incorrect preparation may allow an excessive amount of weld metal to be deposited, resulting in excessive shrinkage which in turn will cause distortion of the fabrication being welded.

What is a fillet weld?

Any fusion weld which is approximately triangular in cross section is a fillet weld. Such a weld may be deposited in an inside corner, an outside corner or a lap joint. See Fig. 50.

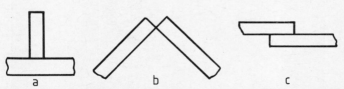

Fig. 50. *Types of fillet weld: (a) inside corner, (b) outside corner and (c) lap joint*

Where are the root and toe of the weld?

These are indicated in Fig. 51.

Fig. 51. Root and toe

What is meant by the throat thickness of a fillet weld?

The throat thickness is the shortest distance from the root to the surface of the weld deposit. See Fig. 52.

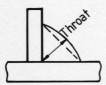

Fig. 52. Throat thickness

What is the leg length of a fillet weld?

This is the distance from the root to the toe of a fillet weld, measured along the fusion face. See Fig. 53.

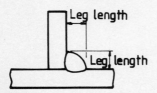

Fig. 53. Leg length

What is the minimum leg length of a fillet weld?

The leg length should be at least as long as the plate thickness. If, however, the plate thicknesses are unequal, the leg length should be at least equal to the thickness of the thinner of the two plates being welded.

If a fillet weld is 10 mm, does this refer to leg length or throat thickness?

The leg length, *not* the throat thickness, should be 10 mm.

How is the leg length related to the throat thickness?

In Fig. 54 the leg lengths are each 10 mm. The length of a diagonal line drawn from the toes is 14.1 mm, the effective throat thickness would be half of this, i.e. 7.05 mm. Therefore, the ratio

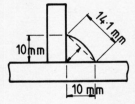

Fig. 54. *Relationship between leg length and throat thickness*

of 14.1 to 10 is constant as far as the diagonal line is concerned. In welding practice, in order to ensure that welds are of sufficient strength, the throat thickness is taken to be 0.8 of the leg length in the case of a fillet weld.

What are the recommended weld joint preparations and their symbols?

These are shown in Fig. 55.

69

Type of Joint	Illustration	Symbol
Flanged butt		⫛
Square butt		‖
Single-V-butt		V
Single bevel butt		ⱽ
Single-V-butt		Y
Single bevel butt with broad root face		ⱼ
Single-U-butt		⋃
Single-J-butt		ⱼ
Backing or sealing run		⌣
Fillet weld		◺

Fig. 55. Recommended joint preparations and their symbols (BS 499, 1980)

How are these symbols applied to engineering drawings?

If the weld symbol is placed above the reference line as in Fig. 56a, the weld should be made from the opposite side of the joint, as shown in Fig. 56b, not the side to which the arrow is pointing.

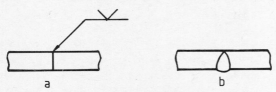

Fig. 56. *(a) Weld symbol. (b) Position of weld*

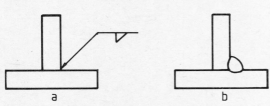

Fig. 57. *(a) Weld symbol. (b) Position of weld*

Where the welding symbol is placed below the reference line, as in Fig. 57a, the weld should be made on the side of the joint indicated by the arrow, as shown in Fig. 57b.

When the weld symbol is shown on each side of the reference line, the joint should be welded from each side.

What weld symbols are used to indicate 'weld all round', 'weld to be made on site' and 'weld to be radiographically examined'?

These symbols are shown in Fig. 58. Fig. 58a indicates that the weld is made continuously, unbroken around the sections prepared for welding. The symbol is placed between the horizontal line and the arrow line.

71

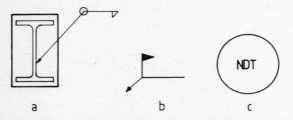

Fig. 58. (a) Weld all round the joint. (b) Weld on site. (c) Weld subject to non-destructive testing

If the weld has to be made on site, a flag is shown in the same position. See Fig. 58*b*.

Where the weld has to be radiographically examined (non-destructive tested), the symbol used can be seen in Fig. 58*c*.

How is a single bevel butt joint indicated on a drawing?

This is shown in Fig. 59. The arrow must point to the plate that is to be bevelled; the symbol below the line indicates that the weld has to be made from the side to which the arrow is pointing.

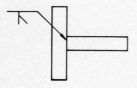

Fig. 59. Single bevel butt joint

How is an intermittent fillet weld indicated?

Fig. 60 is the method of indicating on an engineering drawing that an intermittent fillet weld is required. The 400 mm is to be

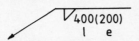

Fig. 60. Method of showing an intermittent fillet weld. ℓ represents area to be welded, e represents area to be unwelded when making weld

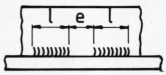

Fig. 61. Completed intermittent fillet weld

welded, whilst the 200 mm shown inside the bracket is the distance left between the welds. Fig. 61 shows the completed weld joint.

How is the welding process to be used indicated on an engineering drawing?

British Standards have adopted a numerical indication system of welding processes for reference purposes. The number of the

Fig. 62. Numerical indication of process

process to be used is shown on an engineering drawing as in Fig. 62.

What are the gas shielded arc process indication numbers?

  13 – Gas shielded metal arc welding.
131 – MIG welding.
135 – MAG welding non-inert gas shield.
136 – Flux-cored metal arc welding with a non-inert gas shield.
  14 – Gas shielded welding with a non-consumable electrode.
141 – TIG welding.
  15 – Plasma arc welding.

73

How is a single vee butt weld with a flush (flat) surface shown?

This can be seen in Fig. 63.

Fig. 63. *Symbol for a single vee butt weld with a flush finish*

9
Principles of the TIG welding process

What is tungsten inert gas welding?

This is an arc welding process in which an arc is struck between the tip of a non-consumable electrode and the work to be welded. The joint edges are melted and fused together by the heat of the arc; a filler rod may be used to provide additional metal to fill the joint. During welding, the molten weld pool and the tungsten electrode are protected from the atmosphere by a shield of inert gas which issues from the nozzle of the welding torch.

What is a non-consumable electrode?

An electrode which is not melted or intentionally used up during the process of welding. It will not become part of the welded joint.

What is an inert gas?

A gas which will not chemically combine with any other known element. Argon and helium are the only inert gases normally used for TIG welding.

What type of electrical current supplies are required at the arc when TIG welding?

Alternating current (a.c.) or direct current (d.c.) may be used, depending on the material being welded.

What is the difference between a.c. and d.c. supplies?

A.C. changes its direction flow in the circuit at twice the frequency of the supply. The polarity of the electrode and the work also change as the direction of the current flow changes. The frequency of the a.c. mains supply is normally 50 hertz, therefore the direction of the current flow and the electrode polarity change 100 times each second. See Fig. 64(*a*).

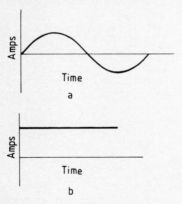

Fig. 64. (a) A.C. supply. (b) D.C. supply

D.C. flows continuously in the same direction through the circuit, and the electrode or work may be connected to either the positive or negative terminal of the welding plant. When TIG welding it is usual to connect the electrode negative. See Fig. 64(*b*).

How will the type of current supplied to the arc affect welding conditions?

If d.c. is used with the electrode connected to the negative, approximately two thirds of the arc heat developed will be at the work (positive) and the remaining one third at the electrode. This condition is suitable for many TIG welding applications, including the welding of stainless steel, copper, nickel, titanium, and their alloys. If the electrode is connected to the positive, the electron flow from the work to the electrode effectively removes refractory oxides from the surface of certain materials. However, this condition is entirely unsatisfactory because two thirds of the heat is at the electrode, causing it to overheat. This results in disintegration of the electrode, which can lead to the presence of tungsten inclusions in the weld. It is for this reason that electrode positive is seldom used. See Fig. 65.

Fig. 65. *(a) Electrode positive. (b) Electrode negative. (c) Electrode with a.c.*

When a.c. is used, the arc heat is developed equally between the electrode and the work being welded. This condition allows satisfactory removal of oxides without overheating the electrode. See Fig. 65. A.C. is used extensively for the welding of materials such as aluminium and magnesium which have refractory oxides.

What is a refractory oxide?

A refractory oxide usually has a higher melting point than that of the metal from which it is formed. For example, pure aluminium has a melting point of 658°C, whilst its oxide alumina melts at 2050°C. Obviously if this refractory oxide is not removed before and during fusion welding, fusion of the joint edges will be severely impeded.

Which metals must be welded with alternating current?

Aluminium, magnesium and their alloys, and also alloys containing appreciable amounts of these metals, such as aluminium bronze, all have a refractory oxide on their surface; even cast iron contains impurities of a refractory nature. These materials are TIG welded using a.c. supplies at the arc in order to remove the oxides and impurities by cathodic cleaning.

What is cathodic cleaning?

This is the removal of refractory oxides during welding. The cleaning half cycle is when the electrode is positive. The alternative half cycle increases the heat input to the work, allowing penetration to be attained. See Fig. 66.

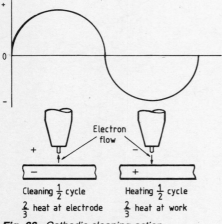

Fig. 66. Cathodic cleaning action

What problems are associated with a.c. welding supplies?

Inherent rectification. This condition exists when an a.c. arc is established between two dissimilar metals, e.g. the tungsten electrode and the aluminium work piece. The current flow in the electrode positive half cycle is less than in the electrode negative half cycle. This results in a loss of oxide removal when the electrode becomes positive, and an excess of current flowing in one direction each time the electrode becomes negative. The latter is known as a d.c. component which causes overheating of the welding transformer.

Partial rectification. With an a.c. arc, the current passes through zero each time it changes its direction of flow in the circuit. The open circuit voltage must be sufficiently high enough to re-ignite the arc. Because a greater voltage is required to re-ignite the positive half cycle, this may be delayed or lost, resulting in a loss of oxide removal.

How are these problems controlled?

Inherent rectification can be overcome by including a d.c. suppressor unit in the circuit (Fig. 67).

Partial rectification can be overcome by using suitable arc re-ignition devices (see page 84).

Fig. 67. *Inherent rectification. Note reduced current flow on the positive half cycle*

What is pulsed TIG welding?

This is a technique requiring a power source which supplies a background current of a relatively low value, with a pulsating current of a higher value superimposed onto the background current.

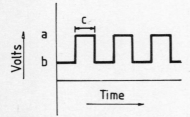

Fig. 68. *Typical wave form when pulsed TIG welding. a = pulsed current, b = background current, and c = pulse duration*

The purpose of the background current is to maintain ionization of the welding arc gap between the pulsations, whilst the pulsed current is used to melt and fuse the joint. The pulsed current may vary in amplitude, duration and frequency (Fig. 68).

What are the applications of pulsed TIG welding?

Pulsed arc welding provides very fine control over the heat input to the workpiece being welded. This allows accurate control over penetration and the depth of fusion, even where there is poor joint alignment and fit-up.

The technique is used extensively when automatic TIG welding. It also has many uses when manual TIG welding, including the welding of thin sections and materials of differing thickness and/or type. It is also used for situations where the heat build-up in the weld area must be kept to a minimum, an example of this being the welding of stainless steels.

Pulsed TIG welding is for d.c. applications only.

10
Equipment for TIG welding

What equipment is required for TIG welding?

For welding with d.c. supplies at the arc, the minimum require-
ment is a d.c. power source, a supply of shielding gas, flowmeter/
regulator, and tungsten electrodes, along with a TIG welding
torch complete with accessories. To assist in striking the arc, a
spark-starter or a high-frequency unit is recommended.

When welding with a.c. supplies at the arc, an a.c. power
source, high-frequency unit, suppressor unit or surge injector,
supply of shielding gas, flowmeter/regulator, tungsten electrodes,
TIG welding torch and accessories are neessary.

In addition to the above items, the following should be
available and can be used with either a.c. or d.c. supplies at the
arc:

- Filler materials (BS 2901).
- Jigs and backing bars (BS 3019).
- Crater-filling device.
- Gas lens.
- Protective clothing.

What type of power source is used when TIG welding?

The power source should provide a drooping characteristic
output. D.C. supplies may be obtained by passing a.c. from the
mains supply through a transformer–rectifier providing d.c.

supplies at the arc. Alternatively, d.c. may be produced by means of a motor generator welding set.

When a.c. is required, this is normally provided by passing the mains supply through a suitable transformer and choke reactor. A suppressor unit is also required in the circuit.

What is a drooping characteristic type of output?

With this type of output, the open circuit voltage available to strike the arc falls to a much lower value immediately the arc is struck. Fig. 69 shows a typical volt/amp curve for this type of

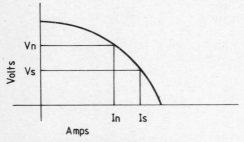

Fig. 69. Drooping characteristic output

welding machine. It can be seen that any variation in arc length causes the arc voltage to vary; this is accompanied by a corresponding variation in the amperage so that the power (wattage) available at the arc (amps × volts = watts) remains fairly constant.

How does the arc power remain constant?

Fig. 69 shows this. Vn is the voltage and In is the amperage for a normally correct arc length. If the arc length shortens, the voltage drops to Vs whilst the amperage increases to Is.

82

What types of welding torches are available for TIG welding?

Lightweight pencil torches may be used with welding currents up to 50 A; other air-cooled torches can handle currents up to 100 A; water-cooled torches are available with current capacities of over 500 A.

How do tungsten electrodes used for d.c. welding differ from those used for a.c. welding?

Electrodes alloyed with a small percentage of thorium are recommended for d.c. welding, as thorium assists in maintaining the point life of the electrode. These thoriated electrodes may also be used for many applications with a.c. welding.

High-quality welds can be attained with a.c. by using tungsten electrodes alloyed with a small percentage of zirconium. The result of using these electrodes is a very reduced degree of

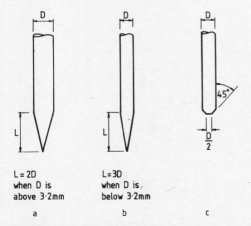

L = 2D
when D is
above 3·2mm

L = 3D
when D is
below 3·2mm

a   b   c

Fig. 70. *Recommended electrode shape when (a, b) using d.c. supplies at the arc, and (c) when using a.c. supplies at the arc*

83

tungsten contamination of the weld metal. Zirconium also improves the emission of electrons from the electrode, giving easier striking and improved arc stability. Electrodes should be preground before use. See Fig. 70.

When is a high-frequency unit or a spark starter used?

The high-frequency unit may be used to initiate the arc when using either a.c. or d.c. supplies. With a.c. the high frequency may be continuous, this assisting in the re-ignition of the arc at the beginning of each half cycle when the current's direction of flow changes.

When welding with d.c. supplies, the high-frequency unit is used only for arc initiation.

The spark starter is used only with d.c. supplies, and is an alternative device to the high-frequency unit for starting the arc.

When is a suppressor unit used?

It is used when TIG welding with a.c. supplies at the arc. The suppressor unit improves arc stability by overcoming the problem of inherent rectification (see Chapter 9).

What is the purpose of the surge injector?

This is an arc re-ignition device which may be used when welding with a.c. supplies at the arc. The surge injector supplies a pulse of around 300 V into the welding circuit at the beginning of each electrode positive half cycle. This prevents a partial rectifying action from occurring (see Chapter 9), and also permits the use of a power source having an open circuit voltage of less than 80 V.

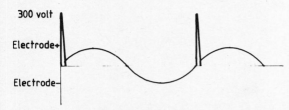

Fig. 71. *Surge injector supplies a 300 V surge at beginning of each electrode positive half cycle*

It should be remembered that when a surge injector is incorporated in the welding circuit, the high-frequency supply is used for arc ignition only. See Fig. 71.

How does the crater-filling device work?

On completing a run of weld, the welder releases a footswitch and the current automatically decays to zero over a period of time. This period can be preset by the welder, and is usually between 1 and 10 seconds. This gradual reduction of current allows the weld crater to be filled at the end of the weld without risk of burn-through. It is often made use of when TIG welding stainless steel, as it reduces the incidence of crater cracks.

How is the shielding gas to the weld zone controlled?

The shielding gas may be supplied from an evaporator for bulk storage users, or from a high-pressure cylinder. The gas is first passed through a gas-pressure regulator, where it is reduced to a suitable working pressure. It is then passed through a flow-meter which controls and indicates the quantity of gas flowing to the weld zone. A gas economiser is usually incorporated to conserve gas when not actually welding.

85

What is a gas lens?

This consists of an attachment to the nozzle of the welding torch which reduces turbulence of the gas as it leaves the nozzle. This permits a greater tungsten stick-out from the nozzle, giving a

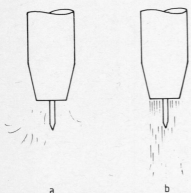

a   b

Fig. 72. *Gas flow (a) without gas lens, (b) with gas lens*

better view of the weld pool and also improved access when welding in corners or deep vee preparations. Welding may now be carried out with a lower gas-flow rate, giving improved gas economy. See Fig. 72.

How is the gas economiser used?

When the welding torch is hung on the gas economiser lever, the gas is shut off. With modern welding machines the gas flow is usually controlled by means of a footswitch.

What filler wires are used when TIG welding?

These are specified in BS 2901, and include:

Part One. Ferritic steels. A15. The filler rod must contain iron base, 0.12 per cent carbon, 0.3 to 0.9 per cent silicon, 0.9 to 1.6

per cent manganese, 0.4 per cent aluminium. Sulphur and phosphorus are impurities usually present, but must not exceed 0.04 per cent.

Part Two. Austenitic stainless steel. A typical filler rod would contain 19 to 22 per cent chromium, 9 to 11 per cent nickel, 1 to 2 per cent manganese, 0.2 to 0.6 per cent silicon, 0.03 per cent carbon. Niobium is added as a stabilizer.

Part Three. Copper. C7. Copper 98.5 per cent plus small additions of lead, aluminium, titanium, manganese, silicon and nickel.

Part Four. Aluminium. GIB. Aluminium 99.5 per cent plus small additions of copper, silicon, iron, manganese and zinc.

Part Five. Nickel. Nickel plus small amounts of titanium.

11
TIG welding techniques

How is the arc initiated when TIG welding?

When using a.c., the high-frequency unit is used. The electrode tip is held about 6 mm from the work. When the high-frequency unit is switched on, the spark produced crosses the arc gap and ionises it. This allows the welding current to flow, forming an arc. As the electrode does not contact the work, no contamination of the electrode tip occurs. The high-frequency unit may be left on when welding with a.c. supplies, as this improves arc stability.

When d.c. is used, the high-frequency unit is used for arc-starting only. An alternative method of striking the arc when using d.c. is the spark starter, which provides a single pulse high-voltage spark to ionise the arc gap. The d.c. arc may also be touch-started onto the work, or on a carbon block which is then transferred to the work. However, this method is not recommended due to the risk of damaging the electrode point.

How far should the tungsten electrode extend from the gas nozzle?

The electrode extension is usually between 3 mm to 8 mm (Fig. 73). A longer electrode extension may be allowed when a gas lens is fitted inside the nozzle.

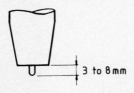

3 to 8 mm

Fig. 73. *Tungsten electrode extension*

What arc length is recommended when TIG welding?

The arc length should approximately be equal to the electrode diameter up to a maximum of 6 mm.

How should the shielding gas flow rate be adjusted?

The shielding gas flow rate depends upon several factors, the main ones being the material being welded, the amount of

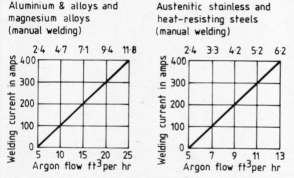

Aluminium & alloys and magnesium alloys (manual welding)

Austenitic stainless and heat-resisting steels (manual welding)

Fig. 74. *Relationship between gas flow rate, welding current and material being used*

welding current and the type of shielding gas. The graphs in Fig. 74 show typical gas flow rates for argon according to the welding current and the type of material being welded.

How should the electrode be shaped when TIG welding?

The recommended shapes are shown in Fig. 70.

What is meant by slope and tilt angles?

These terms refer to the angles of the electrode and filler rod in relation to the surface of the metal being welded and the direction of travel along the weld joint. These can be seen clearly in Fig. 75.

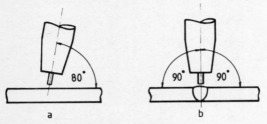

Fig. 75. (a) Slope angle. (b) Tilt angle

How should the welding torch be held in order to maintain the correct angles?

The welding torch is designed to be held in the overhand position when butt welding, and in the underhand position when fillet welding, as shown in Figs. 76 and 77.

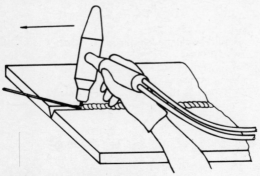

Fig. 76. Overhand position

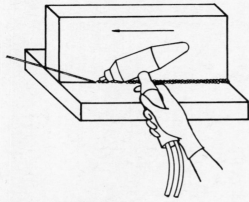

Fig. 77. Underhand position

What is the procedure for carrying out a butt weld on 2 mm thick aluminium in the flat position?

The welding machine should be set to give an a.c. output of between 100 to 120 A. A 3.2 mm diameter zirconiated tungsten electrode should be used along with a 2.4 mm diameter filler rod. The argon flow rate should be adjusted to provide 6 litres per minute through a 12.7 mm bore gas nozzle.

The joint edges should be cleaned and prepared to give a closed square butt joint, which must be tacked or held in a suitable jig. With the torch held overhand, the arc may be struck at the right-hand edge of the joint using the torch and filler rod angles as shown in Fig. 77. As soon as the weld pool is formed, the torch should be moved in a leftwards direction. The speed of travel should be consistent with complete penetration and fusion; this may be judged by observing the width of fusion across the weld pool. During welding, filler metal is added by scratching the filler rod into the leading edge of the weld pool. This ensures that the rod is kept at the same earth potential as the work, reducing the risk of electric shock.

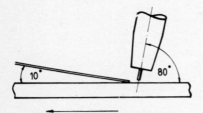

Fig. 78. *Technique of TIG welding*

On completion of the joint, the arc is broken by releasing the footswitch. Allow argon to flow over the completed weld for a few seconds during cooling. See Fig. 78.

What changes are necessary if a similar joint is required in austenitic stainless steel?

The welding machine is set to give d.c. output in the region of 70 to 80 A, the electrode being connected to the negative terminal. A 1.6 mm thoriated tungsten electrode should be used, along with a 2.4 mm stainless steel filler rod of suitable composition. The argon flow rate should be adjusted to provide 3 litres per minute through a 10 mm bore gas nozzle.

How does the technique differ for fillet welding?

The angle of tilt is approximately 45° to each plate. This is best achieved by holding the torch in the underhand position. When fillet welding, a smaller-bore gas nozzle is recommended to enable easier access to the joint. When welding a tee fillet joint, the greater heat loss by conduction through the plates necessitates the use of higher currents than required for butt welding similar thicknesses.

How is vertical welding carried out?

A vertical upwards technique is preferred (Fig. 79). The torch is held in the underhand position, and a weave may be employed to

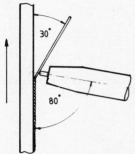

Fig. 79. *Vertical upwards welding*

assist in fusion of the joint edges when welding butt joints or outside-corner joints. Weaving is not normally necessary when fillet welding.

What advantages may be attained from using jigs and backing bars when TIG welding?

The main functions of a jig are holding, locating and backing.

The jig holds the parts in position whilst tacking or welding is done. In many cases, tacking may not be necessary and the jig simply assists in controlling distortion during welding.

Locating assists in quick assembly of the parts in correct relation to each other, thereby ensuring that the fabrications assembled in the jig are of uniform dimensions.

Backing refers to the support given to the underside of the joint. Suitable backing bars help in reducing burn-through and collapse of thin sections, and may also be used to assist in shaping the penetration bead on a butt weld.

93

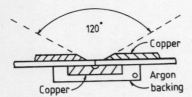
Fig. 80. *Jig with backing bar*

Many jigs used when TIG welding will supply shielding gas to the underside of the joint. This is necessary when welding materials such as stainless steel, which oxidizes rapidly at high temperatures. A jig complete with chills is useful for dissipating heat from the weld area when welding such material. Fig. 80 shows a simple jig and backing bar for use when butt welding.

How may shielding gas be supplied to the inside of a pipe?

This can be achieved by simply allowing the shielding gas to flow along the bore of the pipe. However, improved results are obtained by localizing the shielding gas around the joint area. This is achieved by using a device similar to that shown in Fig. 81.

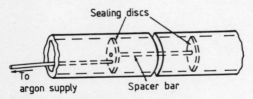
Fig. 81. *Gas backing inside a pipe*

12
MIG and TIG spot welding

What are the MIG and TIG spot welding processes?

These are extensions of the MIG and TIG welding processes, and produce local fusion-type spot welds of consistent size and depth of penetration in a variety of metals and alloys. These include carbon steels, nickel and its alloys. Aluminium and its alloys may also be welded, provided special preparation is carried out.

MIG SPOT WELDING

What equipment is required in addition to standard MIG welding equipment?

The control unit should be fitted with a timer which will switch on and off the welding current, shielding gas and wire feed. A selection of spot-welding nozzles are required to suit the type of

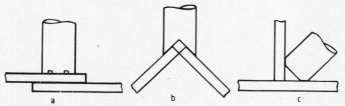

Fig. 82. *MIG spot welding nozzles and joints. (a) lap joint. (b) Outside corner joint. (c) Tee fillet joint*

joint being welded. The purpose of the nozzle is to give the correct contact-tube-to-work distance and to direct the gas shield; it may also be used to apply a moderate amount of pressure to the joint. See Fig. 82.

What are the applications of MIG spot welding?

Welds may be produced on lap joints, the maximum recommended thickness of the top plate being 3 mm. Outside corner joints and tee joints may be spot tack welded using the nozzle shapes described.

Which shielding gases are used?

This depends on the material being welded. For example, nickel and nickel alloys require pure argon; carbon steels use argon + CO_2; and austenitic stainless steels require argon + 1 per cent oxygen.

How is a MIG spot weld carried out?

The contacting joint surfaces must be free of all foreign matter. The welding variables of wire feed speed, voltage, arcing time, gas flow rate and electrode wire diameter are adjusted for the joint being welded. Locate the nozzle on to the spot being welded and

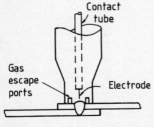

Fig. 83. MIG spot welding

depress the trigger switch to allow the shielding gas to flow, purging the weld area. The wire then feeds and contacts the plate surface and the arc is started. Arcing continues for the preset period of time, then the wire feed stops, followed by the welding current also stopping, allowing the wire to retract inside the nozzle. Gas flows for a short time while the weld cools. The welding cycle is now complete. See Fig. 83.

How does the quality of weld compare with that of a resistance spot weld?

The welds are of high quality, and when lap joints are welded on similar plate thicknesses a small 'pip' on the underside of the plate indicates complete penetration.

What are the advantages of this process over resistance spot welding?

1. Welds may be carried out when access to only one side of the joint is possible.
2. No limitation is placed on the thickness of the bottom plate when lap joints are welded.
3. Outside corner and tee joints may also be welded.
4. Only one side of the joint will be marked after welding.

TIG SPOT WELDING

What equipment is required in addition to standard TIG welding equipment?

A timer is required for the welding cycle, switching on and off the high-frequency spark, welding-current shielding gas and cooling water. It is recommended that a crater-filling device be included in the circuit to reduce the risk of crater cracks. A water-cooled

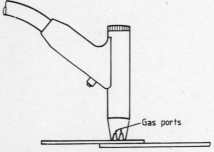

Fig. 84. *TIG spot welding torch*

TIG spot welding torch (Fig. 84) capable of handling up to 400 A is normally used, together with a selection of spot welding nozzles.

What are the applications of TIG spot welding?

These are similar to MIG spot welding; however, tee joints cannot be carried out due to the absence of filler metal. It should be noted that when carbon steels are welded, best results are obtained if killed steel is used, as otherwise porosity may result.

What shielding gas is used?

Pure argon may be used for all applications.

What electrical conditions are required?

Direct current with the electrode connected to the negative terminal.

98

How is a TIG spot weld carried out?

The contacting surfaces are cleaned and the joint set up for welding. The welding variables of gas-flow rate, arcing time, current value and current decay time are adjusted to suit the job. The nozzle is placed on the spot to be welded and the trigger switch depressed. Argon gas begins to flow (pre-purge), then the high-frequency unit and the welding current are switched on. The arc is initiated and arcing continues for the preset period of time. The crater-filling device permits a gradual fall in welding current until the arc is extinguished. Argon gas continues to flow for a short time (post-purge) as the weld and the tungsten electrode cool down. The welding cycle is now complete.

What advantages does TIG spot welding have over other resistance spot welding processes?

These are similar to MIG spot welding. However, due to the absence of any filler material, TIG spot welds will be flush with, or just below the surface of, the top plate. This may, however, be considered an advantage in many circumstances.

13
Plasma arc welding and cutting

What is plasma?

This is a state of gas which has been thermally ionized. In simple terms this means that if a gas is heated to a sufficiently high temperature, it will become capable of easily conducting an electric current. Practical use is made of this principle in the plasma arc welding and cutting processes. The plasma may be produced as a jet in which the work does not form part of the electrical circuit (non-transferred arc), or in the form of a transferred arc in which the work becomes part of the electrical circuit.

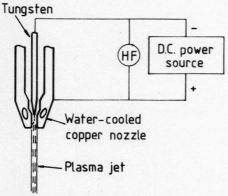

Fig. 85. *Non-transferred plasma jet*

100

How is a non-transferred plasma arc produced?

An electric arc is struck between the central tungsten electrode and the torch nozzle (Fig. 85). The plasma gas is passed through the torch and is restricted by the small orifice of the nozzle, which heats and ionizes it, resulting in a high-temperature, high-velocty jet of plasma.

What are the applications of a plasma jet?

The non-transferred plasma arc jet may be used for thermal metal-spraying operations, and also for the cutting of materials which are electrically non-conductive such as textiles, glass, wood and plastics.

How does the transferred plasma arc operate?

Fig. 86 shows that the work is connected in the electrical circuit. The nozzle of the welding torch is at the same electrical potential as the work, so the current to the nozzle must be limited by a resistor. To start the process, a pilot arc is struck between the

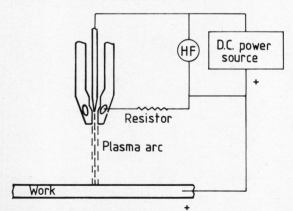

Fig. 86. *Transferred plasma arc*

tungsten electrode and the nozzle, the current flowing through the resistor. As the resulting plasma jet from the nozzle comes into contact with the work it provides a lower resistance path for the current. This allows the arc to transfer from the nozzle to the work, permitting much higher currents to be used.

What are the applications of the transferred plasma arc?

The transferred plasma arc is used for the welding and cutting of a wide range of ferrous and non-ferrous metals.

How is the pilot arc started?

A high-frequency spark unit may be included in the circuit to ionize the gap between the electrode and the nozzle. Once the main arc is established, the high-frequency unit is switched off.

How is atmospheric contamination prevented when plasma arc welding?

A shielding gas nozzle which encircles the plasma nozzle provides a protective gas shield over the weld area (Fig. 87). In most instances the shielding gas is of a similar composition to the plasma gas.

Shielding   Plasma
  gas   gas

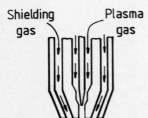

Fig. 87. Shielding and plasma gas

What gas mixtures are used when plasma arc welding?

This depends on the material being welded. The table below lists some of the materials welded and the gases used.

Material	Shielding gas and plasma gas
Stainless steels, mild steel and nickel alloys	Argon-hydrogen mixture
Low-alloy steels, copper and its alloys, titanium and its alloys	Pure argon

What advantages does plasma welding have over TIG welding?

Plasma arc welding offers a number of advantages. Among these are:

1. The electrode cannot come in contact with the weld pool, therefore the risk of tungsten inclusions is reduced and the electrode does not become contaminated.
2. The high current density of the constricted arc gives increased welding speeds, reduced distortion, a smaller heat-affected zone, and depth of penetraton approximately twice that obtained by TIG welding.
3. Arc stability is good even at currents less than 0.1 A. This makes the process suitable for welding extremely thin sections such as foil.
4. The torch stand-off (arc length) is less critical due to the cylindrical shape of the arc. This allows better visibility and access to the joints.
5. Less filler material is required due to deeper penetration and the narrow weld bead which is produced.

How does the plasma gas flow rate affect penetration?

When using low plasma gas flow rates, a soft arc condition is produced which is most suitable for welding thin sections where penetration must be kept to a minimum. The effect of increasing the plasma gas flow rate is to produce a stiffer arc column which can completely penetrate material of above 6 mm thickness. This penetration is achieved by using a 'keyholing' technique. This involves penetrating a small hole at the leading edge of the weld

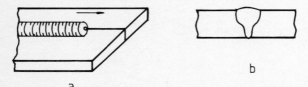

a

b

Fig. 88. Keyholing technique. (a) Note 'keyhole' indicating good penetration. (b) Neat, even penetration attained by using this technique

pool which flows behind the keyhole to form the weld bead shape (Fig. 88). This method uses a very economical amount of filler material; in fact it may often be dispensed with altogether.

How does plasma cutting differ from plasma welding?

When the plasma arc process is used for the cutting of metals, much higher plasma gas flow rates are used producing an extremely high-velocity plasma arc. This is able to melt the material being cut and removes the molten metal by the kinetic energy of the plasma stream. The plasma gases differ from those used for welding, and a secondary gas is often used to improve the efficiency of the cutting process.

Is there any change in the gases used?

Low-cost nitrogen gas or gas mixtures of nitrogen and hydrogen are used for most applications (hydrogen increases the voltage at the arc with any given amperage).

Mixtures of argon and hydrogen are used for producing high-quality cuts; this completely eliminates any nitrogen contamination of the cut edge.

104

What is a secondary gas?

Although some torches rely solely on the plasma gas for cutting, others are designed to use a secondary gas in addition to the plasma gas, to increase cutting efficiency. The secondary gas is

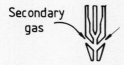

Secondary gas

Fig. 89. *Secondary gas added*

usually carbon dioxide, oxygen or air. These gases are of an oxidizing nature and must therefore be injected into the plasma downstream of the tungsten electrode (Fig. 89).

What are the main applications of plasma arc cutting?

Materials which are difficult or impossible to cut by the oxy-fuel gas cutting process can now be cut. Examples of these are ferrous metals such as stainless steel and cast iron, and non-ferrous metals such as aluminium, magnesium, copper, nickel and their alloys. The highest-quality cuts are obtained within the range of 3 mm to 50 mm thickness. The process is economical for thicknesses up to 150 mm.

How high is the quality of a plasma cut edge?

High-quality cuts with the oxy-fuel gas process are limited to carbon steels. The arc plasma process can produce high-quality cuts in both ferrous and non-ferrous materials. However, the kerf width may be 1½ to 2 times the width of an oxy-fuel gas cut. The sides of the kerf also taper slightly, being wider at the top. One of the cut faces is of a slightly higher quality than the other because

Note taper of
cut face

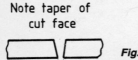

Fig. 90. *With the arc plasma process, one of the cut faces is of slightly higher quality than the other*

the arc locates itself on one side of the cut (Fig. 90), according to the current path through the work. Best results may be obtained by strategic positioning of the positive return lead on the work.

14
Safety precautions

What are the main hazards associated with gas shielded arc processes?

The main types of hazard directly associated with these processes are:

1. Arc-eye and radiation burns.
2. Electric shock.
3. Poisoning or asphyxiation from fumes or shielding gases.
4. Fire or explosion when working in the vicinity of flammable materials.
5. Burns from hot metal and spatter.

What are the effects of arc radiations on the skin and eyes?

Arc rays consist mainly of three types of radiation:

1. Intensity of light, which in itself does not cause permanent injury, but if not reduced results in eye strain.
2. Infra-red or heat radiation, which may be damaging but is not normally a hazard as the immediate feeling of discomfort caused to the welder will result in self-protective measures being taken.
3. Ultra-violet rays. These present a real problem as their damaging effects are not felt at the time of exposure but some hours

later. Even a very brief exposure to ultra-violet radiation may cause an extremely painful condition known as arc-eye or welding flash. The skin may also be damaged by these rays, resulting in a condition similar to severe sunburn.

What precautions should be taken against the effects of arc radiations?

Before welding, the welder must protect himself by wearing suitable protective clothing such as dark long-sleeved overalls, leather apron and welding gloves. The eyes and face should be protected by wearing a suitable head-shield fitted with the correct shade of filter lens to BS 679 (see table opposite).

People working in the vicinity of arc welding or cutting operations must be screened from arc radiation, and they should also be provided with a pair of anti-flash spectacles.

Walls in welding workshops should be painted with a non-reflective matt paint to reduce the effect of reflected arc radiations.

What is the correct shade of filter lens?

This depends upon (a) the value of current being used, (b) the process being used, and (c) the surrounding level of light.

The shade and recommended uses (BS 679) are shown in the table opposite.

Is the risk of an electric shock greater with some processes than with others?

Yes; of the processes described in this book, MIG welding may be considered the safest. This is because the open-circuit voltage rarely exceeds 50 V. TIG welding may use open circuit voltages up to 110 V and some plasma-cutting machines may require up to 200 V open circuit.

Filters recommended for welding

Welding process	Approximate range of welding current (in amperes)	Filter(s) required
Metal-arc welding (coated electrodes) Continuous covered-electrode welding Carbon dioxide shielded continuous covered-electrode welding	Up to 100	8/EW 9/EW
	100–300	10/EW 11/EW
	Over 300	12/EW 13/EW 14/EW
Metal-arc welding (bare wire) Carbon-arc welding Inert-gas metal-arc welding Atomic hydrogen welding	Up to 200	10/EW 11/EW
	Over 200	12/EW 13/EW 14/EW
Automatic carbon dioxide shielded metal-arc welding (bare wire)	Over 500	15/EW 16/EW
Inert-gas tungsten-arc welding	Up to 15	8/EW
	15–75	9/EW
	75–100	10/EW
	100–200	11/EW
	200–250	12/EW
	250–300	13/EW 14/EW

Where two or more shade numbers are recommended for a particular process and current range, the higher shade numbers should be used for welding in dark surroundings and the lower shade numbers for welding in bright daylight out of doors

The use of arc-ignition devices, such as a high-frequency spark generator, may increase the shock risk even further.

Why are d.c. supplies considered safer than a.c. supplies?

When d.c. is used, the open-circuit voltage required to strike the arc is normally less than that when using a.c. Moreover, when a shock is received from an a.c. supply there may be difficulty in letting go of a live conductor. This does not normally occur when one receives a shock from a d.c. supply.

What precautions should be taken to reduce the risk of electric shock?

Some of these precautions are:

1. All electrical equipment should be in first-class condition, and all connections and cable insulation inspected regularly.
2. All machinery should be correctly earthed. It is also important that the work being welded is at earth potential; this may be accomplished by earthing the work itself, or by earthing the return lead at its source.
3. When working in a confined space, damp or cramped positions, the risk of receiving an electric shock will be increased. Consideration should be given to the use of a wooden duckboard or rubber mat to insulate the welder from earth. Protective clothing and rubber-soled boots also provide a degree of insulation.
4. A no-load low voltage device is available which keeps the open-circuit voltage down to a minimum until the arc is struck.
5. The welding lead and the return lead must be of a sufficient cross section to carry the total welding current without overheating.
6. When working at a height or in an insecure position, even a mild electric shock could result in a fall, so under such circumstances wearing a safety harness is recommended.

What hazards are represented by the use of shielding gases when welding?

Shielding gases such as argon, helium, nitrogen and carbon dioxide are not classed as toxic, but may be dangerous when working in a confined space. These gases may displace the breathable air in the space, resulting in asphyxiation of the welder.

What precautions must be taken to reduce the risk of asphyxiation?

With the exception of helium, the shielding gases used with these processes are heavier than air. This means that, when welding in a confined space, they tend to build up from the bottom of (say) a

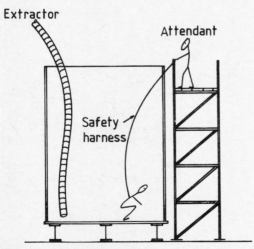

Fig. 91. When working in a confined space, the presence of an attendant and the wearing of a safety harness are advisable

111

vessel. Where possible these gases should be allowed to pass through a manhole at the bottom of such a vessel. If this is not possible they must be removed by an extractor (Fig. 91).

Helium, being lighter than air, may present the reverse of the aforementioned problem when working in a vessel with an enclosed top.

Are any other precautions advisable when welding in a confined space?

It is advisable to have a responsible person in attendance outside the vessel, and a safety harness should be worn to assist in removing the welder from the confined space should difficulties arise (Fig. 91).

What toxic fumes may be encountered during welding?

Many toxic fumes may arise from the surface of the material being welded. Among these are galvanized steel, cadmium plating, chromium plating, paint and also plastic coatings.

Oil and grease from the surface of the metal being welded may also present a hazard. It is important that consideration be given to fume extraction at all times.

How can the risk from these fumes be minimised?

An area 25 mm wide either side of the weld preparation should have the surface coating removed. Oil and grease can be removed by a suitable degreasing procedure.

The use of local extraction is recommended to remove these undesirable fumes at their source; extractor attachments are available which fit to the gun when MIG welding (Fig. 92).

What aids to breathing are available for welders' use?

Particularly when welding in a confined space, some form of respirator should be used. This may take the form of: (1) A dust respirator for protection against particulate fumes. (2) An air-feed welding helmet into which clean compressed air is fed at a slightly

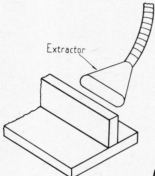

Fig. 92. *Local extraction*

positive pressure, thus preventing the fumes from entering the welding helmet. This also provides a much cooler environment for the welder. (3) Compressed-air breathing apparatus. This provides total protection from toxic fumes. For further information on this, see BS 4275 1974 (Recommendation for the selection, use and maintenance of respiratory protective equipment).

Do degreasing agents present a fume hazard?

Certain degreasing solvents produce toxic phosgene gas when exposed to arc rays (ultra-violet). For this reason arc welding should not be carried out within 30 metres of a degreasing plant. Any work which has been degreased must not be taken into the welding area until thoroughly dry.

What circumstances can give rise to a fire or explosion?

The use of welding or cutting equipment in the vicinity of flammable materials, or welding vessels which have contained flammable materials.

Care must be exercised when handling cylinders of compressed gas. For example, the use of oxygen or a flammable gas in a confined space presents a high degree of risk should the recommended precautions not be adhered to.

What precautions should be observed to reduce the risk of a fire?

1. Where possible all flammable materials should be removed from the immediate welding area. If this is impractical they should be covered with an asbestos sheet as protection against sparks.
2. Fire-fighting equipment of the correct type should always be available.
3. In particularly hazardous situations, fire watchers should always be available.

How should a vessel which has contained a flammable substance be made safe before welding?

The two main methods of rendering a vessel safe are:

1. Complete removal of the flammable substance.
2. Displacement of all air space within the vessel to prevent the build-up of any explosive mixture.

How can a vessel having contained a flammable substance be rendered safe?

This may be carried out by steaming, i.e. using low-pressure steam. This should be continued for at least half an hour, but may

114

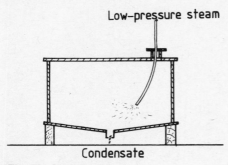

Fig. 93. *Steaming*

in some cases require 2 to 3 hours, depending on the size of the vessel or its contents. During steaming the condensate must be permitted to drain away. See Fig. 93.

Large vessels may require a longer steaming period, and must be certified safe by a competent person before welding.

How can the air space in a vessel be displaced?

This is achieved by filling the vessel with a material which will not form an explosive mixture, e.g. nitrogen foam, inert gas, sand or

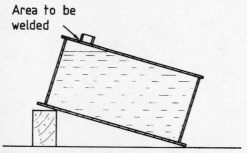

Fig. 94. *Small vessel filled with water to displace air*

115

water. Sand and water are often suitable for small vessels or tanks (Fig. 94).

How are burns caused when gas shielded arc welding?

Burns may be caused by flying sparks and molten metal from the welding arc, by flying slag when chipping welds deposited with a flux-cored electrode wire, by arc radiations or by touching the hot welded parts.

How may the risk of burns from hot metal and slag be reduced?

This is a matter of using the correct equipment and also wearing the correct protective clothing. Particular care must be taken to avoid falling sparks or hot metal when welding or cutting in the overhead position, or in a confined space.

What protective clothing is available?

The welder's hands should be protected by wearing welding gloves; the body by using long-sleeved overalls, leather armlets, apron, leggings, spats and boots. The head may be protected during positional welding by wearing a leather skull cap. When chipping slag or wire-brushing, it is extremely important that protective goggles are worn.

Index

QUESTIONS & ANSWERS

Gas Welding and Cutting
P H M Bourbousson

All the facets of successful welding are covered in the form of a well-chosen series of questions and answers that get progressively more and more involved with the subject. Has been revised and updated, with new chapters on the control of distortion and on flame brazing of aluminium.
144 pages   0 408 01180 7

Pipework and Pipe Welding
L J Rose

An introduction to the types of pipeline in common use, methods of pipeline assembly and the testing of joints. It will be informative and helpful to pipe fitters, maintenance engineers, and all concerned in the installation, maintenance and repair of pipelines, ranging from domestic ones to the highly specialised systems used in machine tool operation and control.
110 pages   0 408 00108 9

Electric Arc Welding
3rd EDITION **K Leake and N J Henthorne**

A practical introduction to electric arc welding, which is today the most widely used of all welding processes. The information is presented in question and answer form in a progressive sequence, which provides a practical course of instruction from the simple building-up of pads of weld metal to advanced welding techniques. This third edition takes account of the revised British Standards on the classification of electrodes and on weld symbols.
144 pages   0 408 01128 9

Newnes Technical Books
Borough Green, Sevenoaks, Kent TN15 8PH